# Wi-Fi 6

## Protocol and Network

For a complete listing of titles in the
*Artech House Mobile Communications Library,*
turn to the back of this book.

# Wi-Fi 6

## Protocol and Network

Susinder R. Gulasekaran

Sundar G. Sankaran

ARTECH HOUSE

BOSTON | LONDON

artechhouse.com

**Library of Congress Cataloging-in-Publication Data**
A catalog record for this book is available from the U.S. Library of Congress.

**British Library Cataloguing in Publication Data**
A catalogue record for this book is available from the British Library.

Cover design by Charlene Stevens

ISBN 13: 978-1-63081-842-5


**685 Canton Street**
**Norwood, MA 02062**



10 9 8 7 6 5 4 3 2 1

*Dedicated to all those who have contributed to Wi-Fi and love Wi-Fi*

# Contents

Acknowledgments xiii

Chapter 1 Introduction 1
1.1 IEEE Standardization Process 2
1.2 Wi-Fi Alliance 3
1.3 Spectrum 5
1.4 Outline of Chapters 5

Chapter 2 Wi-Fi 1 to 5 Overview 7
2.1 MAC Layer Overview 8
2.1.1 Network Architecture 8
2.1.2 MAC Frame Format and Types 10
2.1.3 Medium Access Rules 12
2.1.4 AP Discovery Process 18
2.1.5 Connection Establishment and Termination 19
2.1.6 Off-Channel Scanning 20
2.1.7 Quality of Service 21
2.1.8 Fragmentation and Aggregation 23
2.1.9 STA Power Save Methods 26
2.1.10 Multicast Traffic 32
2.1.11 STA Roaming 32
2.1.12 AP Channel Switching 34
2.2 PHY Layer Overview 34
2.2.1 Wi-Fi Generation 1 (802.11b) 35
2.2.2 Wi-Fi Generation 2 (802.11a) 37

2.2.3 Wi-Fi Generation 3 (802.11g) 39
2.2.4 Wi-Fi Generation 4 (802.11n) 39
2.2.5 Wi-Fi Generation 5 (802.11ac) 44
2.3 References 49

Chapter 3 Wi-Fi 6 51
3.1 HE PPDU Formats 52
3.1.1 Long OFDM Symbol 53
3.1.2 GI and HE LTF 54
3.2 Single User Operation 54
3.2.1 Coding and Modulation 54
3.2.2 Dual Carrier Modulation 55
3.2.3 HE TxBf and Channel Sounding 55
3.2.4 Midamble 57
3.2.5 Multi-TID AMPDU 58
3.2.6 Blockack Length 58
3.2.7 Aggregation Limits 59
3.2.8 Duration-based RTS-CTS 59
3.3 Multiuser Operation 60
3.3.1 OFDMA Resource Unit Allocation 60
3.3.2 Downlink MU Operation 61
3.3.3 Uplink MU Operation 63
3.3.4 Trigger Frame Types 68
3.3.5 Buffer Status Report 71
3.3.6 MU EDCA 74
3.4 Target Wake Time 75
3.4.1 TWT Agreement Setup 76
3.4.2 Types of TWT Agreements 77
3.4.3 TWT Summary 78
3.5 BSS Coloring 79
3.5.1 BSS Color Collision 80
3.5.2 BSS Color Change Announcement 81
3.6 Spatial Reuse 81
3.6.1 OBSS PD-based SR 81
3.6.2 PSR-based SR 84
3.6.3 Practical Challenges in SR 85
3.7 Enhanced Multi-BSSID Advertisement 86
3.7.1 Profile Periodicity 88

3.8 Preamble puncturing 88
3.9 Operating Mode Indication 89
3.10 HE Capability Advertisement 90
3.11 When to do SU, MU-MIMO and MU-OFDMA 90
3.11.1 Examples 93
3.12 References 101

Chapter 4 Wi-Fi Security 103
4.1 Wired Equivalent Privacy 104
4.1.1 WEP Authentication 104
4.1.2 WEP Encryption 104
4.1.3 WEP Vulnerabilities 105
4.2 Wi-Fi Protected Access 106
4.2.1 WPA Personal 106
4.2.2 WPA Enterprise 108
4.2.3 WPA Encryption 109
4.2.4 Message Integrity Check 109
4.3 WPA2 110
4.3.1 WPA2 Authentication 110
4.3.2 WPA2 Encryption 110
4.3.3 Message Integrity Check 111
4.3.4 PMK Caching 111
4.3.5 Opportunistic Key Caching 112
4.3.6 Fast BSS Transition 112
4.3.7 Protected Management Frames 112
4.3.8 Wi-Fi Protected Setup 113
4.4 Key Reinstallation Attack 114
4.5 WPA3 114
4.5.1 WPA3 Authentication 115
4.5.2 WPA3 Encryption 117
4.5.3 WPA3 Transition Mode 117
4.5.4 Device Provisioning Protocol 117
4.5.5 Opportunistic Wireless Encryption 118
4.6 Multichannel Man-in-the-Middle 119
4.6.1 Operation Channel Validation 120
4.6.2 Beacon Protection 121
4.7 Fragmentation and Aggregation Attacks 121
4.8 Rogue AP Detection 122

4.9 Summary 123
4.10 References 124

Chapter 5 Wi-Fi Implementation 127
5.1 Hardware 127
5.1.1 Radio Requirements 129
5.1.2 Host Processor 136
5.1.3 Memory Size 137
5.2 Software 138
5.2.1 Rate Adaptation 138
5.2.2 Scheduler 140
5.2.3 Buffer management 141
5.2.4 Packet Processing Speed 142
5.2.5 Channel Selection 143
5.2.6 STA Roaming Algorithm 144
5.2.7 STA Power Save Algorithm 145
5.2.8 Capability Advertisements 145
5.3 Regulatory Requirements 146
5.3.1 Allowed Channel Frequencies 146
5.3.2 In-Band Power and Out-of-Band Emission Limits 147
5.3.3 Dynamic Frequency Selection 150
5.4 Wi-Fi Certifications 152
5.5 References 155

Chapter 6 Wi-Fi 6E 157
6.1 6-GHz Channelization 158
6.2 AP Discovery 158
6.2.1 Minimizing Probe Traffic 159
6.2.2 Out-of-Band Discovery of a 6-GHz AP 160
6.2.3 Fast Passive Scanning of a 6-GHz Only AP 161
6.3 EMA support in 6 GHz 161
6.4 Security Modes in 6 GHz 162
6.5 Beacon Advertisement and PHY Rates in 6 GHz 162
6.6 On Channel Tunneling 163
6.7 Fast Initial Link Setup 164
6.7.1 FILS and Fast BSS Transition 166
6.8 6-GHz Regulatory Requirements 167
6.8.1 FCC 167
6.8.2 ETSI 170

6.9 Wi-Fi 6E Certification 170
6.10 Wi-Fi 6E Implementation Challenges 170
6.11 160-MHz Multiple AP Deployment in 6 GHz 171
6.12 References 173

Chapter 7 Wi-Fi Deployment 175
7.1 Residential Wi-Fi Deployment 175
7.1.1 Mesh 176
7.1.2 AP Placement 177
7.1.3 Channel Bandwidth Selection 177
7.2 Enterprise Indoor Wi-Fi Deployment 178
7.2.1 Power over Ethernet 178
7.2.2 Choosing AP Specification 179
7.2.3 Channel Bandwidth Selection 180
7.2.4 Estimating the Number of APs 183
7.2.5 Mitigating Inter-AP Interference 189
7.2.6 Upgrading Existing Deployment 190
7.2.7 Reducing Overhead in High-Density Deployments 191
7.3 Outdoor Wi-Fi Deployment 191
7.4 Post Deployment Site Survey 192
7.5 References 193

Chapter 8 5G Overview and Wi-Fi Comparison 195
8.1 Evolution of Cellular Technologies (1G-4G) 197
8.1.1 First Generation Cellular Technologies 197
8.1.2 Second Generation Cellular Technologies 197
8.1.3 Third Generation Cellular Technologies 198
8.1.4 Fourth Generation Cellular Technologies 199
8.2 Fifth Generation Cellular Technology 199
8.3 Comparison with Wi-Fi 203
8.3.1 Technology 203
8.3.2 Security 204
8.3.3 Ease of Deployment 204
8.3.4 Total Cost of Ownership 204
8.3.5 Use Cases 205
8.4 Summary 206
8.5 References 207

Chapter 9 Conclusion 209

List of Acronyms and Abbreviations 211

About the Authors 221

Index 223

# Acknowledgments

Our sincere thanks to the reviewers for their time and prompt feedback to improve the book. We are thankful to our Wi-Fi industry friends and colleagues who have helped enrich our Wi-Fi knowledge over many years. A special thanks to Dr. Srikanth Subramanian, Nanocell Networks Pvt. Ltd., and to Jason Wang, Commscope Inc., for their constructive suggestions to enhance the content and value of this book. Finally, this book would not have been possible without the support and encouragement from our family members especially during the challenging times of a pandemic.

# Chapter 1

# Introduction

Wi-Fi has become a necessity in today's world. It has evolved into a fundamental utility, like electricity and water, that everyone—without any thought—expects to be available everywhere irrespective of whether it is a school, a hotel, a restaurant, a stadium, an office building, or even public transportation.

Wi-Fi is a wireless local area network (WLAN) technology, essentially an Ethernet replacement that allows devices to connect to the internet without being tethered by any wires or cables. It is based on Institute of Electrical and Electronic Engineers (IEEE) 802.11 standards. The absence of wires and cables extends access to places where the wires and cables cannot reach. This also brings down the network deployment cost since it avoids any trenching and drilling needed to create physical connections.

Wi-Fi is one of the greatest success stories of the modern technology era. Wi-Fi enabled devices can connect to the internet via a wireless access point (AP). APs have a range between 10 and 30 meters indoors, and the range can be much longer outdoors. Hundreds of millions of Wi-Fi access points connect billions of computers, smartphones, smart TVs, game consoles, cameras, printers, Internet of Things (IoT) devices, and other consumer devices to the internet to enable millions of applications to reach everyone, everywhere. According to a study commissioned by Wi-Fi Alliance (WFA), this generates a global economic value estimated at $3.3 trillion USD. This is expected to grow to $4.9 trillion by 2025.

The total global mobile traffic is estimated to increase from around 51 exabytes (51 billion gigabytes) per month by the end of 2020 to 226 exabytes per month in 2026. Operators have long been leveraging Wi-Fi networks to scale capacity to meet these ever-increasing bandwidth needs of their subscribers.

One of the main catalysts for the widespread adoption of Wi-Fi is the decision by regulatory bodies around the world to make significant blocks of spectrum available on a license-exempt basis. Unlike licensed bands, where the wireless communication protocol to be used is mandated as part of the licensing process, no protocols were specified for these unlicensed bands. This unregulated environment created opportunities for the industry to develop novel wireless communication protocols for various applications, including computer-to-computer communication. However, this resulted in several companies developing WLAN products that were incompatible with each other. The IEEE 802.11 standard was later developed in order to guarantee compatibility and interoperability.

## 1.1 IEEE STANDARDIZATION PROCESS

IEEE standardization starts with an idea for a standard. A group of individuals or companies form a study group under the IEEE Standards Association's umbrella for up to six months before their work gets approved as an IEEE standards project. The most famous among IEEE standards projects is the IEEE 802 project that focuses on local area network (LAN) standards. Working groups are formed within a project to develop standards to address a particular aspect of the problem. The working groups are assigned a number, which is written after the decimal point for the corresponding projects. WLAN happens to be the eleventh working group formed under IEEE 802, and hence got the number 802.11. The working group opens a PAR – a Project Authorization Request – that includes items such as the purpose and scope of the standard, who will chair the working group, and the expected duration of the project. After the PAR is approved by the IEEE SA's standards board, the working group begins the work to develop the standard. The standard that the working group develops is called a draft standard. At least 75 percent of the sponsor ballot group comprising interested entities or individuals needs to approve the draft standard by a letter ballot before it gets sent to the IEEE-SA Standards Board for approval. Once approved, it becomes an officially ratified standard. Then the publishing process begins, and the standard is made available to everyone.

As the standard gets used and as technology advances, an inevitable need to update the standard arises. This restarts the standardization process all over again. Task groups are formed within working groups to amend or add to the standard and are assigned alphabet(s) beneath the working group. The task group then submits a PAR. Once the PAR is approved, the draft to revise the standard is developed; the draft is then balloted and sent to the IEEE-SA Standards Board for approval.

The document produced by the task group is identified by the project number and working group number, followed by the alphabet(s) assigned to the task group. Task Group a (TGa) is the first WLAN task group formed and the standard developed by this task group is called IEEE 802.11a specification. Table 1.1 lists all the specifications, also known as amendments, developed by the IEEE 802.11 task groups.

Periodically, once every four years or so in the case of IEEE 802.11, a task group is formed to roll up the amendments into the parent specification. The first such task group TGma was formed in 2003 to create a single document that merged 8 amendments (802.11a, b, d, e, g, h, i, j) with the original IEEE 802.11 standard. The resultant document, approved in 2007, became the new base standard named IEEE 802.11-2007. The most recent rolled out standard is IEEE 802.11-2020, which was produced by the fourth maintenance task group referred to as TGmd. Maintenance/revision for IEEE 802.11-2020 is being handled by TGme.

## 1.2 WI-FI ALLIANCE

IEEE 802.11 is a long and complex standard. Despite the best efforts, some sections of the standard are bound to be ambiguous or not fully defined. Furthermore, the standard includes several optional features and different manufacturers might make different choices in their designs. To avoid interoperability problems, a non profit organization called Wireless Ethernet Compatibility Alliance (WECA) was formed by a group of major manufacturers two years after the original standard was ratified. The main objectives of WECA were to promote the use and implementation of the new standard and to certify products that conform to subsets of the 802.11 standard for interoperability. In 2000, WECA changed the name of its technology from Wireless Ethernet to Wi-Fi, a name coined by the brand consulting firm Interbrand as a pun on hi-fi and the organization was renamed WFA.

Since its founding in 1999, WFA has developed several interoperability certification programs to ensure that the Wi-Fi devices from different vendors interoperate seamlessly. These certification programs test various aspects of the 802.11 specifications on devices under test primarily for interoperability. The devices that pass these tests are certified to be "Wi-Fi Certified TM" for the specific 802.11 technology. Wi-Fi Certified TM is an internationally recognized certificate of approval for products indicating that these products have met industry standards of interoperability and security. Currently, WFA has more than 750 companies from around the

**Table 1.1**

IEEE 802.11 Amendments

| *Amendment* | *Year Ratified* | *Description* |
|---|---|---|
| 802.11b | 1999 | OFDM PHY in 5 GHz (Wi-Fi 2) |
| 802.11a | 1999 | 2.4 GHz CCK 5.5 and 11 Mbps (Wi-Fi 1) |
| 802.11d | 2001 | Tx power and frequency compliance for regulatory domains |
| 802.11g | 2003 | OFDM PHY in 2.4 GHz (Wi-Fi 3) |
| 802.11h | 2003 | DFS and TPC to protect radar |
| 802.11i | 2004 | Wi-Fi security enhancements (AES CCMP) |
| 802.11j | 2004 | Japan regulatory domain rules in 4.9-5 GHz band |
| 802.11e | 2005 | QoS enhancements EDCA, UAPSD, blockacks, direct link setup |
| 802.11k | 2008 | Radio resource management |
| 802.11r | 2008 | Fast roaming for WPA2 |
| 802.11y | 2008 | High power operation in 3650 to 3700 MHz |
| 802.11n | 2009 | Wi-Fi 4 |
| 802.11w | 2009 | Protected management frames |
| 802.11p | 2010 | Wireless access for vehicular applications using OFDM in upper 5-GHz band |
| 802.11z | 2010 | Tunneled Direct link setup |
| 802.11v | 2011 | Wireless network management and BSS transition management |
| 802.11u | 2011 | Seamless roaming between participating Wi-Fi networks (Hotspot 2.0) |
| 802.11s | 2011 | Wireless mesh networking |
| 802.11ae | 2012 | Medium access prioritization of management frames |
| 802.11ad | 2012 | 60 GHz Wi-Fi |
| 802.11ac | 2013 | Wi-Fi 5 |
| 802.11af | 2014 | Wi-Fi operation in TV White spaces frequency band (59-790 MHz) |
| 802.11mc | 2015 | Maintenance standard which also added Wi-Fi positioning |
| 802.11ah | 2016 | Standard for Wi-Fi operation below 1 GHz (755-928 MHz) intended for long range and IoT applications |
| 802.11ai | 2016 | Fast Initial Link Setup |
| 802.11aj | 2018 | Amendments to 802.11ad for operation in China |
| 802.11ak | 2018 | Defines how clients with both 802.11 Wi-Fi and 802.3 wired capability can bridge the two mediums and also provides services like VLAN, audio-visual stream, and time synchronization for audio-visual streams |
| 802.11aq | 2018 | Defines how APs can advertise network service information to clients prior to connecting to a network |
| 802.11ax | 2021 | Wi-Fi 6 |
| TGay | To be ratified | Next generation standard upgrade to 802.11ad for 60 GHz |
| TGaz | To be ratified | Next generation Wi-Fi positioning |
| TGbe | To be ratified | Wi-Fi 7 |

world as members and has certified more than 50,000 products for various Wi-Fi technologies.

The Wi-Fi standards, as described earlier, are identified as IEEE 802.11n, IEEE 802.11ac, IEEE 802.11ax, and so on. These are not very user-friendly, so, the WFA introduced a new naming convention in 2018 to easily recognize the Wi-Fi capability of devices. The new Wi-Fi naming scheme easily identifies the different generations by using a numerical sequence, each one of which corresponds to a major advancement in Wi-Fi technology. This numbering scheme makes it easier to see the progression of different generations of Wi-Fi. It is not obvious to most people that IEEE 802.11ac was the next generation from IEEE 802.11n, but it is much easier to see that Wi-Fi 6 is the next development on from Wi-Fi 5. This approach is very similar to the concept of technology generations, like 2G, 3G, 4G, and 5G, used in the cellular industry.

## 1.3 SPECTRUM

Since Wi-Fi operates in unlicensed spectrum, Wi-Fi networks can be deployed by anyone without any license or permission from the government. In the early days, Wi-Fi used the unlicensed 2.4 GHz ISM band. However, the 2.4-GHz band is crowded as it is used by more than just Wi-Fi. Old cordless phones, baby monitors, microwave ovens, and Bluetooth devices tend to use the 2.4-GHz band. To overcome the connectivity issues such as slow and dropped connections, Wi-Fi devices started using the 5-GHz unlicensed band also known as Unlicensed National Information Infrastructure (U-NII) band. Initially, Wi-Fi was allowed to operate only in a small fraction of the U-NII band. But as the demand for Wi-Fi increased, to mitigate congestion, regulatory bodies opened up more parts of the U-NII band. Table 1.2 lists various sub bands of the U-NII band. The availability as well as the regulations for using the allowed sub bands varies depending on the country.

## 1.4 OUTLINE OF CHAPTERS

Wi-Fi has gone through six generations over the last 25 years. Each new generation is built on the previous one to achieve improvements in speed and spectral efficiency. Over these six generations, Wi-Fi speed and efficiency have improved by three orders of magnitude.

This book is about the latest sixth-generation Wi-Fi, based on 802.11ax standard. Wi-Fi 6 supports a maximum data rate of nearly 10 Gbps with spectral

**Table 1.2**
U-NII Bands

| *Band* | *Frequency Range (MHz)* | *Bandwidth (MHz)* |
|---|---|---|
| U-NII-1 | 5150–5250 | 100 |
| U-NII-2A | 5250–5350 | 100 |
| U-NII-2B | 5350–5470 | 120 |
| U-NII-2C | 5470–5725 | 255 |
| U-NII-3 | 5725–5850 | 125 |
| U-NII-4 | 5850–5925 | 75 |
| U-NII-5 | 5925–6425 | 500 |
| U-NII-6 | 6425–6525 | 100 |
| U-NII-7 | 6525–6875 | 350 |
| U-NII-8 | 6875–7125 | 250 |

efficiency reaching 62.5 bps/Hz allowing concurrent transmissions to/from as many as 74 devices. In addition to improving data rate and spectral efficiency, Wi-Fi 6 incorporates new technologies to improve the Wi-Fi performance in congested areas with a lot of connected devices and to extend the battery life of Wi-Fi enabled devices.

With the background on the first five generations of Wi-Fi in Chapter 2, Chapter 3 describes the key new features of Wi-Fi 6. Security is a key component of any wireless system. Wi-Fi security is presented in Chapter 4. Chapter 5 presents the implementation aspects of a Wi-Fi 6 system. With the regulatory agencies around the world recently opening up the 6-GHz spectrum for unlicensed use, a new class of Wi-Fi devices called Wi-Fi 6E devices that extend Wi-Fi 6 into the 6-GHz spectrum is being launched. The novel aspects of Wi-Fi 6E are presented in Chapter 6. Chapter 7 summarizes the factors one needs consider while deploying Wi-Fi networks, while Chapter 8 compares the latest generation of Wi-Fi and cellular technologies. Chapter 9 concludes the book with a preview into what is in the future for Wi-Fi.

# Chapter 2

# Wi-Fi 1 to 5 Overview

This chapter provides an overview of the first five generations of Wi-Fi technologies. Wi-Fi protocols [1] are best described as two layers: medium access control (MAC) layer and physical (PHY) layer. The MAC layer governs medium access by enforcing a set of rules that dictate how to access the medium to transmit or receive data. In addition to this, the MAC layer is responsible for the management of all connections as well as power save management. The PHY layer converts data bits received from the MAC layer into a radio frequency (RF) waveform for transmission and performs the reverse process of converting received RF waveform into bits that are sent to the MAC layer. The term Wi-Fi generation is associated with the PHY layer. Although every generation of PHY layer technology brought some associated MAC layer changes, several MAC layer aspects and features were developed completely independent of the PHY layer as part of Institute of Electrical and Electronic Engineers (IEEE) 802.11e, h, i, k, r, v, and ai. The IEEE 802.11 standard ratified in 1997 was the first to define the foundations of Wi-Fi MAC layer along with two PHY layers based on direct sequence spread spectrum (DSSS) and frequency hopping spread spectrum (FHSS). However, not many devices adopted it until the 802.11b standard arrived in 1999. The first generation of Wi-Fi refers to devices that operate in the 2.4-GHz band supporting the 802.11b standard. The second generation of Wi-Fi refers to devices that operate in the 5-GHz band using orthogonal frequency division multiplexing (OFDM) technology as defined by the 802.11a standard. Wi-Fi generation 3 refers to the 802.11g standard that adopted the 802.11a OFDM technology for operation in the 2.4-GHz band incorporating modifications needed for coexistence with 802.11b devices. Devices supporting 802.11b, 802.11a or 802.11g standards are often referred to as legacy devices. Wi-Fi generation 4 refers to the 802.11n standard that supports operation in both 2.4-GHz and 5-GHz bands and devices supporting Wi-Fi 4 are called high throughput (HT) devices. Wi-Fi generation 5 is based on the 802.11ac standard for 5-GHz operation and devices supporting it are known as very high throughput (VHT) devices.

## 2.1 MAC LAYER OVERVIEW

The Wi-Fi MAC layer is designed to be highly flexible with several options to choose from depending on the network architecture and use case. This section provides an overview of the Wi-Fi MAC layer and describes the most common options that are relevant to residential and enterprise deployments.

### 2.1.1 Network Architecture

There are four types of Wi-Fi network topology or architecture as listed below and illustrated in Figure 2.1:

1. Infrastructure or basic service set (BSS);
2. Ad hoc or peer-to-peer or independent BSS (IBSS);
3. Wireless distribution system (WDS);
4. Mesh BSS (MBSS).

Infrastructure or BSS is the commonly used network architecture in which all devices connect and communicate through a single device called the access point (AP). The AP administers network service to all devices connected to it, commonly referred to as clients or stations (STAs). The AP is connected via a backhaul link to the internet and often has an integrated router capability. Any data traffic between STAs has to be routed through the AP. In the ad hoc network architecture, devices can directly connect and communicate with each other without any centralized entity. The wireless distribution system (WDS) architecture provides wireless backhaul link to APs located in places where it is not possible to provide wired backhaul link. It helps extend the wireless range of a network using multiple hops over wireless. The mesh BSS (MBSS) architecture employs the 802.11s standard to form multiple hop mesh networks. The 802.11s mesh is more sophisticated than the WDS architecture and enables formation of complex mesh networks with multiple wireless backhaul links for fault tolerance.

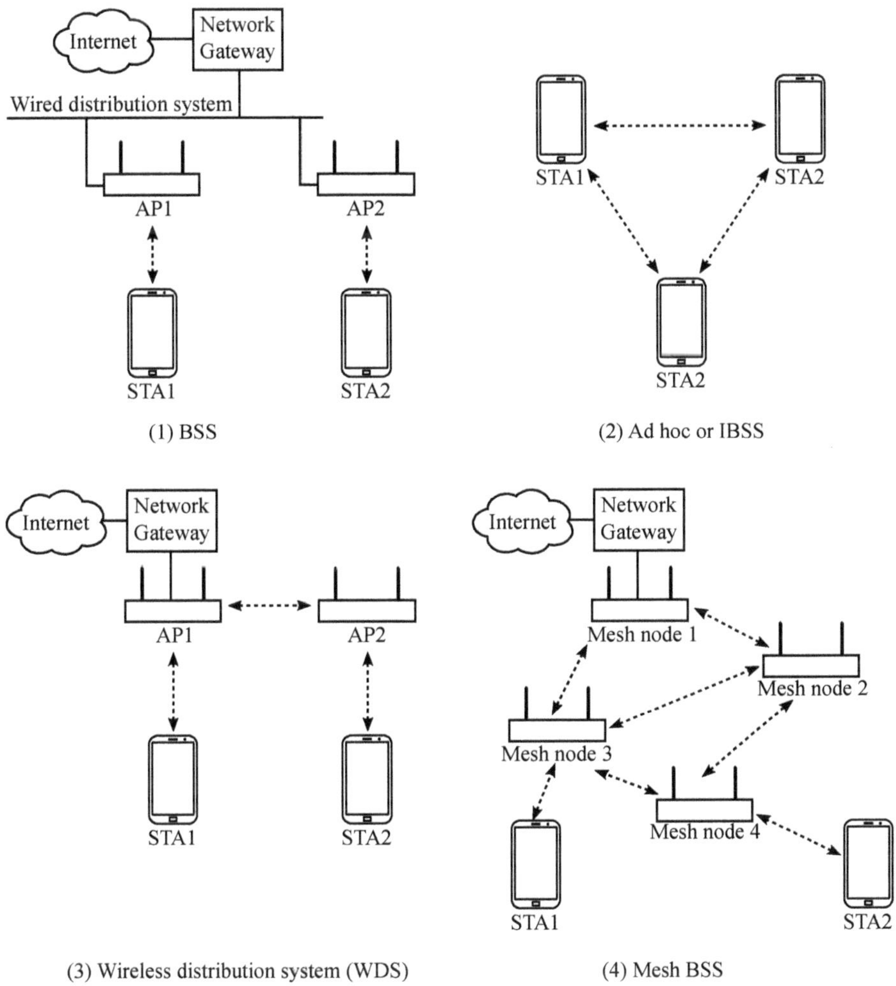


**Figure 2.1** Wi-Fi network architectures.

### 2.1.2 MAC Frame Format and Types

The MAC layer sends the data to be transmitted in a particular frame format, as shown in Figure 2.2, to the PHY layer. There are three components in any MAC frame: MAC header, MAC frame body and frame check sequence (FCS). The MAC frame is also known as MAC protocol data unit (MPDU). The MAC frame body is the MAC payload or data to be transmitted and is also referred as MAC service data unit (MSDU). The MAC header contains information such as frame type and addresses needed to send the packet to the desired destination. The FCS field is a 32-bit cyclic redundancy check (CRC) computed using both header and frame body contents to detect any errors introduced during transit.

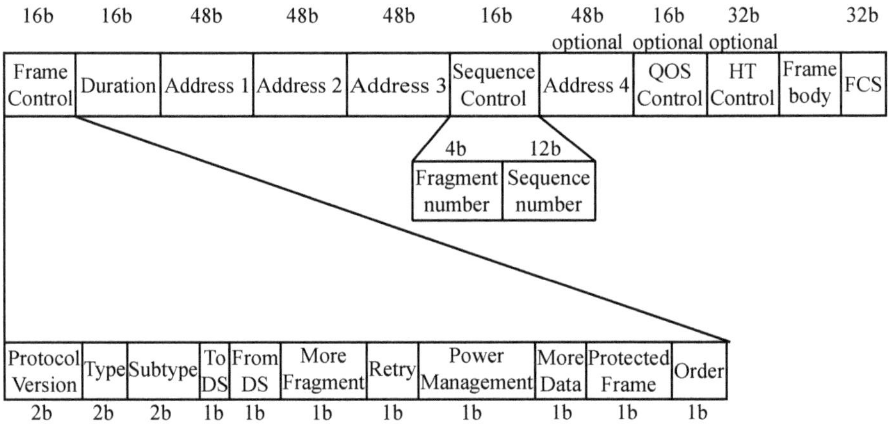


**Figure 2.2** Format of MPDU or MAC frame.

Every Wi-Fi device in a network has its own unique 48-bit (6-byte) MAC address. The Address 1 and Address 2 fields in the MAC header are populated with receiver address (RA) and transmitter address (TA), respectively. RA is the MAC address of the device that is intended to receive the frame, while TA is the MAC address of the transmitting device. The Address 3 and Address 4 fields are assigned to one of BSS identifier (BSSID), source address (SA), or destination address (DA) depending on the network architecture and direction of frame flow as described in Figure 2.3. SA is the MAC address of the device that originated the frame contents, DA indicates the MAC address of the frame's intended destination device, and BSSID is defined as the AP's MAC address. Note in Figure 2.3 that the To DS and From DS subfields of the frame control field indicate the direction of

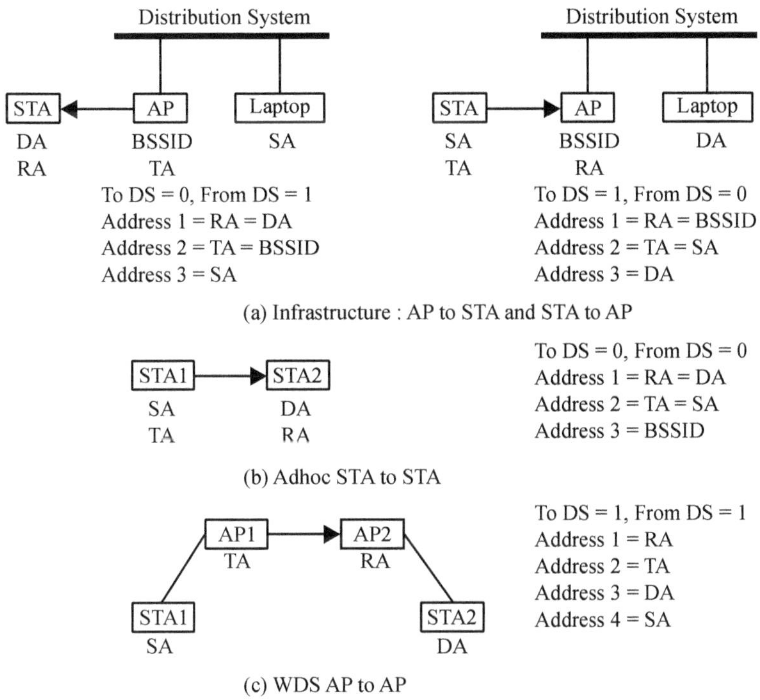


**Figure 2.3** MAC frame addressing.

frame flow (i.e., whether the frame is going from AP to STA or from STA to AP). The frame control field has many important uses such as to indicate frame type, whether the payload is encrypted or not, whether the frame is being retried or not, and power save management. The duration field is used as the network allocation vector (NAV) field for medium reservation. The sequence control field denotes the sequence number for the frame and fragmentation information. The sequence number is an important subfield used to identify duplicate frames and to assemble frames in order before sending them to the higher layers. The Address 4 field is present only in WDS network architecture and the QoS control field is present only in frames that require specific quality of service (QoS). The frame format used in MBSS has an additional mesh header field after the QoS control field that includes two more address fields to facilitate multiple hop routing of frames.

There are three types of MAC frames: data frames, management frames, and control frames. The frame type can be inferred from the type and subtype subfields of the frame control field in the MAC header. Some common and important frame types are described in Table 2.1. While data frames are meant to carry data, management frames are used for connection management and control frames are employed for acknowledgment and for triggering specific actions. Control frames have a smaller MAC header and no payload data as shown in Figure 2.4. Management frames have a structured payload composed of some fixed length fields and one or more variable length fields called information elements (IEs).

| 16b | 16b | 48b | 48b | 32b |
|---|---|---|---|---|
| Frame Control | Duration | RA | TA | FCS |

(a) RTS frame

| 16b | 16b | 48b | 32b |
|---|---|---|---|
| Frame Control | Duration | RA | FCS |

(b) ACK and CTS frame

**Figure 2.4** Control frame format.

### 2.1.3 Medium Access Rules

The rules of Wi-Fi medium access are described by the distributed coordination function (DCF) protocol that is built upon the carrier sense multiple access collision detection (CSMA-CD) protocol used in IEEE 802.3 Ethernet MAC. Some changes were made to address the practical limitations of wireless transceivers and the broadcast nature of wireless medium. The transceiver in Wi-Fi is half-duplex, while it is full-duplex in the case of Ethernet. Half-duplex refers to the constraint of the transceiver to either be in transmit or receive mode at any time, but not both simultaneously. If full-duplex is attempted, the receive circuitry will be overwhelmed because of the low isolation (typically 40 dB) between transmit and receive circuitry. Collision detection is not possible in Wi-Fi owing to the half-duplex constraint.

The DCF protocol is similar to how humans converse in a group, where every individual would first listen to check if someone else is talking, wait for the person talking to finish, and then start talking after a brief random pause. DCF is a distributed or decentralized approach to multiple access. Wi-Fi requires the receiving device to transmit an acknowledgment control frame (ACK) at short interframe spacing (SIFS) time duration from the completion of frame reception, so that the transmitting device can infer a collision event if it doesn't receive the

**Table 2.1**
Common MAC Frame Types

| *Type* | *Subtype* | *Frame Type* | *Purpose* |
|---|---|---|---|
| Data (10) | 0000 | Data | Carry data without quality of service. |
| | 0100 | Null | Use certain fields of MAC header. Carries no data. |
| | 1000 | QoS data | Carry data requiring QoS. |
| | 1100 | QoS null | Use QoS control field of MAC header. |
| Management (00) | 1000 | Beacon | AP capability advertisement, STA synchronization, power management. |
| | 0100 | Probe request | Active scanning of nearby APs. |
| | 0101 | Probe response | Inform STA with AP's capabilities. |
| | 0000 | Association request | For STA to request association with AP. |
| | 0001 | Association response | For AP to indicate association status to STA. |
| | 0010 | Reassociation request | STA uses to change elements of current association or roam from one AP to another. |
| | 0011 | Reassociation response | Used by AP to approve a reassociation request. |
| | 1010 | Disassociation | Used by AP or STA to terminate an association. |
| | 1011 | Authentication | Used between AP and STA for the AP to identify a STA. |
| | 1100 | Deauthentication | Used by AP or STA to terminate a connection. |
| | 1101 | Action | To notify or initiate actions. |
| | 1110 | Action no ack | To notify actions without requiring acknowledgment. |
| Control (01) | 1101 | ACK | Acknowledge unicast data or management frames. |
| | 1000 | Blockack request (BAR) | To request blockack. |
| | 1001 | Blockack (BA) | Acknowledge a sequence of frames. |
| | 1011 | RTS | Prevent collisions from devices close to transmitting device. |
| | 1101 | CTS | Prevent collisions from devices close to receiving device. |
| | 1010 | PS poll | For power save STA to request AP to send one buffered MSDU. |

ACK frame. DCF protocol is best described as a transition among 6 different states as shown in Figure 2.5.

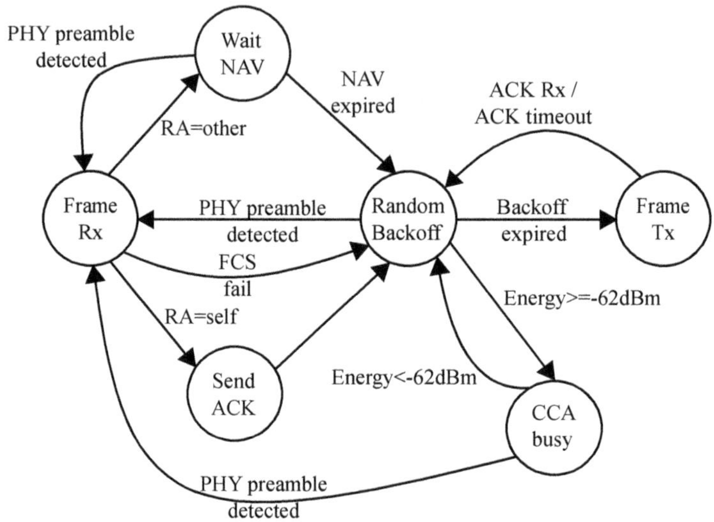


**Figure 2.5** DCF protocol.

A device starts in the random backoff (RBO) state, where the radio is in receive (Rx) mode. As a general rule, the device is always in Rx mode while it is not transmitting. During receive, it is continuously searching for the presence of Wi-Fi PHY preamble, which is a unique signature pattern present at the beginning of any Wi-Fi frame. The device will be in this RBO state for a time duration of DCF interframe space (DIFS) followed by B · ST, where ST is a constant time duration referred to as slot time. Here, DIFS is SIFS + 2 · ST, and B is a random integer picked uniformly between 0 and contention window (CW). The parameter CW is initialized to a value of $CW_{min}$. The value of SIFS is 16 $\mu$s in the 5-GHz band, while it is 10 $\mu$s in the 2.4-GHz band. The ST value is 9 $\mu$s in the absence of 802.11b devices, and 20 $\mu$s in the presence of 802.11b devices for compatibility reason. The random component B · ST is implemented as a backoff counter that starts from B and decrements by 1 for every ST time interval. The time duration in RBO state could be interrupted anytime if the receiver detects a PHY preamble or if the receiver senses any energy>=-62 dBm. In both cases, the current backoff counter value is frozen, say at value $B_f$. In the case of energy>=-62 dBm, the state transitions to clear channel assessment (CCA) busy state and returns back to RBO state when the energy falls below -62 dBm. In the case of PHY preamble detection, the state transitions to frame Rx state in which the receiver tries to decode the type of Wi-Fi frame and verifies the FCS field of MAC header. If the FCS fails, the state transitions back to RBO state. If the FCS verification passes and if the MAC frame is addressed to the device itself (i.e., the RA field matches its own MAC address), then the state transitions to send ACK state in which the device waits for SIFS duration and then transmits an ACK frame with the RA field set to the TA. After the completion of ACK frame transmission, the state transitions to RBO state. If the MAC frame is destined to someone else, then a virtual carrier sense mechanism is adopted by setting what is called a NAV value, which is set equal to the duration field in MAC header. Then the state transitions to wait NAV state, where the radio is in receive mode for NAV $\mu s$ duration. The virtual carrier sense mechanism thus prevents anyone receiving this MAC frame from transmitting on the medium for the next NAV $\mu s$. This is often employed as a technique for reserving a contention free medium access period with the maximum NAV value being limited to 32767 $\mu s$. Typically most data frames set NAV as SIFS + ACK frame duration to protect the ACK frame from any collisions. Upon NAV time expiry, the state transitions to RBO state. Whenever the state transitions from send ACK or CCA busy or wait NAV to RBO state, the duration in the RBO state is set to be DIFS + $B_f \cdot$ ST, i.e., the backoff counter resumes decrementing from where it left. When the state transitions from frame Rx to RBO due to FCS failure, the RBO duration is instead EIFS + $B_f \cdot$ ST, where extended interframe space (EIFS) = SIFS + DIFS + ACK frame duration. This exception for FCS failure event minimizes time wasted due to false PHY preamble detection events. Upon successful RBO duration expiry, the state transitions to frame Tx state in which the device is allowed to transmit any MAC frame. If the transmitted frame is a unicast frame, the device waits to receive an ACK frame from the receiving device in SIFS time. If the ACK frame is not received within ACK timeout period (typically set to SIFS + ACK duration), the state transitions to RBO state and the CW value is doubled. For each consecutive ACK timeout, the CW value is again doubled, up to the maximum value defined by $CW_{max}$. The ACK timeout event could happen when the transmission of two devices collide or the receiving device encounters FCS failure or the ACK frame faces collision. If the ACK frame is received successfully, CW is reset to $CW_{min}$ and the state transitions to RBO state. This exponential backoff mechanism reduces the chance of repetitive collisions. Figure 2.6 illustrates DCF operation.

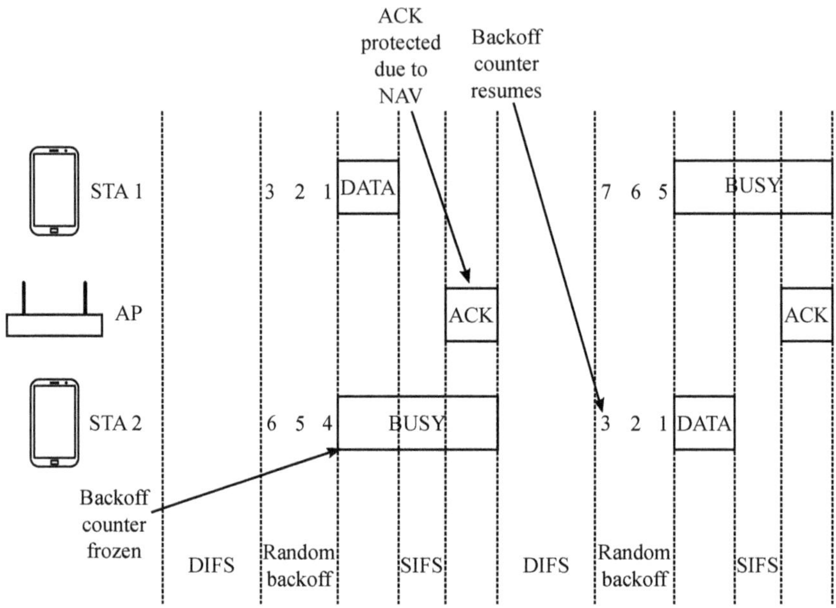


**Figure 2.6** DCF operation.

### 2.1.3.1 Hidden Node Problem and Collision Avoidance

Collision can happen when two devices have the same RBO duration or when the distance between two devices is outside their hearing range. The second scenario is commonly referred to as the hidden node problem. Assume two devices STA1 and STA2 are trying to send a frame to the AP at the same time. As shown in Figure 2.7, the relative positions of the devices is such that the AP is within the hearing range of both STA1 and STA2, but STA1 and STA2 can't hear each other. So, the transmissions from STA1 and STA2 would collide with each other and the AP would not be able to decode the frame.

Wi-Fi has a provision to avoid data frame collision using RTS-CTS frame exchange prior to data frame transmission. Using this provision, STA1 and STA2 contend for the medium and whoever secures medium first sends a control frame called request to send (RTS) with RA field matching the AP's MAC address. The duration field in the RTS reserves the medium for a duration of SIFS+CTS+SIFS+ Data + SIFS + ACK. This RTS frame prevents collisions from devices within the hearing range of the transmitter. In response to RTS, the AP sends a clear to send (CTS) control frame with the duration in CTS frame = duration in RTS frame − SIFS − CTS. The CTS frame clears the medium of collisions from devices within range of the receiver. Though this technique protects data frame from collision, it cannot protect the RTS frame from collision. However, retransmitting an RTS frame takes less time than that for a data frame. If the airtime occupied by data frames are large, and hidden nodes occur frequently, RTS-CTS becomes worthwhile despite its overhead. Due to this collision protection mechanism, the multiple access scheme of Wi-Fi is sometimes referred as carrier sense multiple access collision avoidance (CSMA-CA).

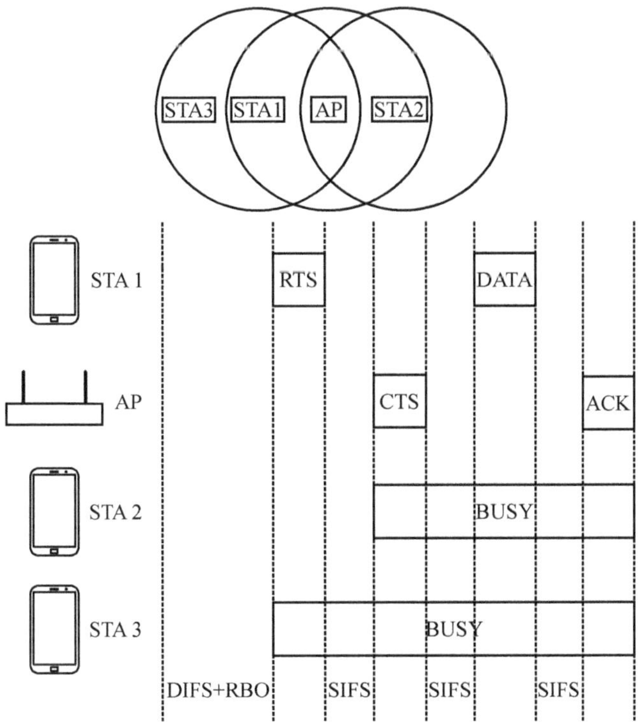


**Figure 2.7** Hidden node scenario.

### 2.1.4 AP Discovery Process

Before a STA can connect to an AP, it must first discover APs in its vicinity along with each AP's channel, BSSID, and capability information. STA performs this discovery by scanning all the allowed channels. For example, in the United States, there are 11 channels in the 2.4-GHz band and 25 channels in the 5-GHz band. There are two types of scanning: passive scanning and active scanning.

During a passive scan, a STA listens on each channel for beacon frames sent periodically by APs. Beacon frames are management frames, broadcasting the BSS's network name or service set identifier (SSID), BSSID, capability information, power management and synchronization information, sent every beacon interval or target beacon transmission time (TBTT). A network is commonly identified by its SSID. The beacon interval is specified as a field in the beacon frame and its typical value is 100 TUs, where 1 time unit (TU) = 1024 $\mu$s. In a multiple AP network deployment such as an enterprise office, all APs advertise the same SSID but each AP advertises a unique BSSID in its beacon. The RA field in a frame's MAC header is set to all ones (or ff:ff:ff:ff:ff:ff in hexadecimal format) to indicate that it is a broadcast frame intended to be received by all devices.

During active scan, STA goes to each channel, broadcasts a management frame called probe request frame with the SSID field set to "wildcard" and listens for a unicast response frame called probe response from APs serving on that channel. The probe response frame also contains AP's SSID, BSSID and capability information. When the SSID field of probe request frame is specified as wildcard, all APs regardless of network name would respond with a probe response frame. If the SSID field is specified, then only APs advertising that specific SSID would respond. A STA typically specifies the SSID field when it wants to discover APs belonging to a known network to which the STA has connected earlier. Active scan is faster than a passive scan but it consumes more power. Excessive active scanning drains the client's battery faster. Also, United States regulatory requirements don't permit active scan on some 5-GHz channels.

#### 2.1.4.1 Scanning Virtual APs

Sometimes, one physical AP can be configured to function as multiple virtual APs. This is also known as multiple BSS configuration. Each virtual access point (VAP) has its own SSID, BSSID, security configuration and it performs everything that a physical AP would do, such as beaconing every TBTT, send probe response for each VAP, and so forth. There are several use cases that require multiple VAPs such as providing different network access permissions or QoS for each VAP. For example, in a home network, one may not want to share Wi-Fi password information with visiting guests, so they can create a separate VAP for guest internet access with different security credentials. In a university, it is desirable to have different access privileges for teachers and students. With multiple VAPs, the wireless management traffic on the network undergoes a multiplicative effect. If there are five physical APs in a building floor on the same channel with eight VAPs configured on each AP, there would be 40 beacon frames on the channel every 102.4 ms (TBTT) and 40 probe response frames for every broadcast wildcard probe request frame from a client.

### 2.1.5 Connection Establishment and Termination

After the STA completes scanning all channels, it decides which AP to connect based on several factors such as user configuration, receive signal strength indicator (RSSI) metric that is a measure of the power level of received frame, medium congestion level on the AP's channel, and AP's capabilities (supported data rates, security, etc). Once the STA determines the best AP to connect to, it sends an authentication request frame to the AP identifying itself and requesting the AP to authenticate. The AP sends ACK for authentication request and responds with an authentication response frame to the STA indicating approval or rejection of the request. The AP usually accepts the request unless the STA MAC address is blacklisted by user configuration. Upon successful authentication, the STA acknowledges the authentication response and sends association request frame containing information on STA capabilities. The AP acknowledges reception of association request frame and checks if the STA capabilities intersect with AP capabilities and whether it is eligible to join. The AP then responds with association response frame to the STA containing a status code which indicates success or failure of association, an association identifier (AID) that is a unique STA identifier assigned by the AP, and AP's capability information. The AID is henceforth used by the AP to reference this STA, and has applications in power save and other features. The capability information is contained in fixed length fields and variable length IEs that are included in beacon, probe request, probe response, association request, and association response frames. Table 2.2 provides a brief description of some common capability information fields and IEs.

Figure 2.8 shows the various steps in connection establishment. After successful association, if the BSS has security enabled then there are a few more frame exchanges to authenticate the STA and install encryption keys, which is described in

**Table 2.2**

Common Capability Information

| *IE or Field Name* | *Description* |
|---|---|
| SSID | Network name. |
| Supported Rates | Indicates supported legacy PHY rates. A separate bit is allocated for AP to indicate the basic rate (lowest PHY rate) of the network. |
| Capability Info | Support for QoS, short slot time, UAPSD power management. |
| HT Capabilities | HT capability for LDPC, Short GI, STBC, delayed BA, AMSDU and AMPDU aggregation limits, supported HT MCS, and number of spatial streams. |
| HT Operation | Indicates the primary channel number and secondary channel offset. |
| VHT Capabilities | VHT capability for LDPC, Short GI, STBC, SU/MU Beamformer and Beamformee, AMSDU and AMPDU aggregation limits, supported bandwidths, supported spatial streams, and maximum MCS for different spatial streams. |
| VHT Operation | Indicates BSS channel width and channel center frequency. |
| Extended Capabilities | Augments the capabilities specified in the capability info field. |
| WMM | STA EDCA parameters AIFSN, $CW_{min}$, $CW_{max}$, TxOP for BE, BK, VI, VO access categories. |
| RSN | Supported cipher algorithms and authentication methods. |
| Country | Defines the country of operation, allowed channels and maximum transmit power. |
| 20/40 BSS Coexistence | Used for 20/40 MHz BSS coexistence. |
| Mobility Domain | Indicate AP's support for 802.11r fast bss transition. |

detail in Chapter 4. An AP or STA can terminate a connection at anytime by notifying the other with either a disassociation frame or deauthenticate frame containing a reason code that specifies the reason for termination.

### 2.1.6 Off-Channel Scanning

Even after establishing connection with an AP, most clients continue to scan all channels intermittently to assess if there is a better AP to connect to. This is usually referred to as off-channel scanning. The channel on which the STA is connected to the AP is called the home channel and any channel other than the home channel is called an off-channel. Off-channel scanning works differently than the scanning process for initial connection. Off-channel scanning is not done continuously as the STA must return to its home channel to receive beacon frame every TBTT. Typically, STAs go off-channel for a brief 20-50 ms duration between beacons to

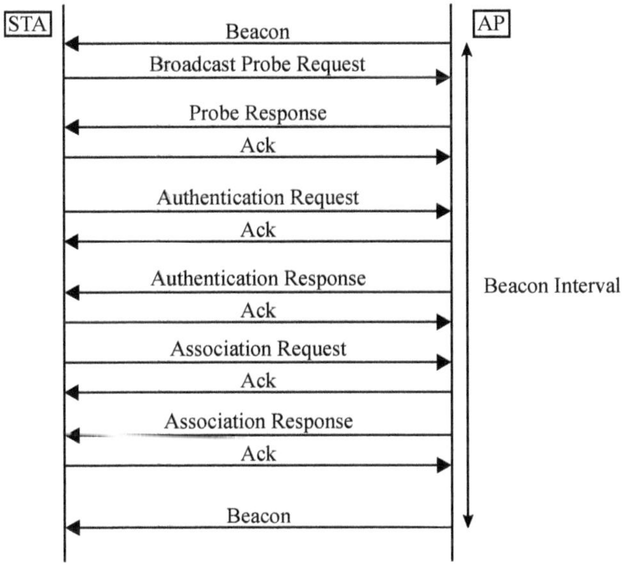


**Figure 2.8** Process of STA association.

scan a different channel. This process is repeated over multiple beacon intervals to cover all the channels. The STA stays connected during off-channel scanning since the fraction of time spent off-channel is very low. Some STA implementations indicate to the AP that they are in power save state before going off-channel and then notify the AP that they are awake upon returning to the home channel. Off-channel scanning enables STAs to learn their RF environment and AP neighbors, which facilitates roaming. More details on STA roaming are provided in Section 2.1.11.

Some AP implementations, especially in enterprise deployments, also perform off-channel scanning to learn their neighbors and RF environment. This information is used for various purposes such as home channel change, wireless security, and also to influence clients to connect to the best AP.

### 2.1.7 Quality of Service

In 2004, WFA started Wi-Fi multimedia (WMM) certification based on the 802.11e standard that adds ability to provide different QoS for different traffic types. WMM replaced the DCF protocol with enhanced distributed channel access (EDCA) protocol. While WMM actually does not guarantee QoS, it provides a way to classify and prioritize traffic of QoS data frame type. This is done by assigning a traffic identifier (TID) from one of eight different values to each QoS data frame. The TID value is specified in the QoS control field of MAC header. The TID assignment is done either based on priority code point (PCP) field of 802.1Q tagged Ethernet frame or by some frame inspection algorithm that detects the application generating these packets. For example, when running a video call application over Wi-Fi on a smartphone, a TID value of 5 is assigned to uplink frames from the smartphone to the AP. Note that downlink Ethernet frames coming from the internet (video streaming for example) to the AP typically do not have any 802.1Q tags. For each of these eight TIDs, there are separate queues to buffer frames and independent MAC sequence number assignment is performed. These eight TIDs are mapped to four access categories: best effort (BE), background (BK), video (VI) and voice (VO). An independent CSMA-CA is implemented for each access category (AC) traffic. The DIFS parameter is replaced with arbitrary interframe spacing(AIFS) $=$ SIFS $+$ AIFSN $\cdot$ ST where AIFSN is dependent on the AC. For each AC traffic, different sets of EDCA parameter (AIFSN, $CW_{min}$, $CW_{max}$) values are assigned to prioritize traffic. Table 2.3 shows the mapping of TIDs to AC and the default EDCA parameters for each AC. In the event of internal collision(two ACs encountering backoff expiry at the same time), exponential backoff happens similar to an external collision event.

**Table 2.3**

Default EDCA Parameters for Different TIDs

| *Priority* | *TID or PCP* | *AC* | *AIFSN* | $CW_{min}$ | $CW_{max}$ |
|---|---|---|---|---|---|
| Lowest | 1, 2 | BK | 7 | 15 | 1023 |
| | 0, 3 | BE | 3 | 15 | 1023 |
| | 4, 5 | VI | 2 | 7 | 15 |
| Highest | 6, 7 | VO | 2 | 3 | 7 |

The EDCA parameters are assigned such that higher-priority TIDs have lower average medium access latency and higher probability of gaining medium access when there are frames for multiple ACs contending for the medium. However, the order of medium access grant does not strictly follow the priority. The $CW_{min}$ for BE and VO are 15 and 3, respectively. There is a higher probability that a random number chosen between 0 and $CW_{min}$ for BE is larger than that of VO, but it is also possible that the random number chosen for BE is smaller than that of VO, granting BE earlier medium access over VO. Over a large number of instances, however, VO traffic experiences less backoff and hence receives more frequent access to the medium compared to BE traffic. To give an example, assume the random slot time component of backoff B happens to be 0 for BE and 2 for VO, then the BE backoff expiry would happen in $\text{SIFS}+3\cdot\text{ST}+0$ $\mu$s while VO backoff expiry would happen later at $\text{SIFS} + 2 \cdot \text{ST} + 2 \cdot \text{ST}$ $\mu$s, thereby granting medium access to BE first over VO. The EDCA parameters to be followed by all STAs is decided by the AP and advertised to all STAs using the WMM field in the beacon frame. The AP is not mandated to use the default EDCA parameters and has flexibility to change them depending on the deployment scenario and use case.

### 2.1.8 Fragmentation and Aggregation

The MAC payload also referred as MSDU can either be fragmented or aggregated, which could be beneficial in certain cases.

#### 2.1.8.1 Fragmentation

Similar to maximum transmission unit (MTU) in Ethernet, there is a fragmentation threshold (in unit of bytes) in Wi-Fi and any MSDU larger than fragmentation threshold is fragmented. A MAC header and FCS are then added to each fragmented MSDU. The more fragment subfield in frame control field of MAC header is set to 1 for all fragmented MSDUs except the last one and the fragment number is indicated in the sequence control field of MAC header. Fragmentation lowers the efficiency of medium usage but in some environments with high collision probability, this may be useful as it lowers the retransmission penalty.

#### 2.1.8.2 Aggregation

With every new generation of Wi-Fi, the PHY data rate increases, resulting in reduction of the PHY payload airtime, while the duration of the PHY header either becomes longer for backward compatibility or remains the same. In Wi-Fi 4, aggregation was introduced for the first time in the MAC layer to increase the ratio of PHY payload airtime to the PPDU airtime, thereby improving the efficiency significantly. There are two levels of aggregation available: aggregate MAC service data unit (AMSDU), which is then encapsulated by aggregated MAC protocol data unit (AMPDU), as shown in Figure 2.9.

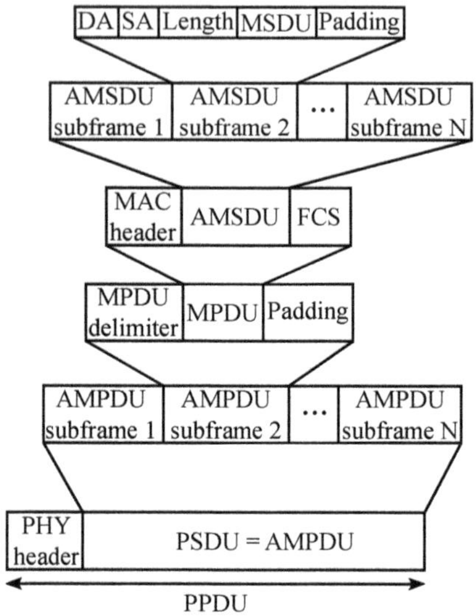


**Figure 2.9** AMSDU and AMPDU aggregation.

An AMSDU is an aggregation of multiple MSDUs. An AMSDU subframe header containing the DA, SA and MSDU length information is added as a prefix to each MSDU to enable de-aggregation at the receiver. Finally, a MAC header and FCS field are added to the AMSDU to create a MPDU or MAC frame. Since the FCS field is present only at MPDU level, the receiver cannot acknowledge individual MSDUs and the entire MPDU has to be retransmitted in the event of FCS failure.

An AMPDU is an aggregation of multiple MPDUs which becomes the payload for the PHY layer known as physical layer convergence procedure (PLCP) service data unit (PSDU). Finally, the PHY layer adds PHY header and tail bits to physical layer conformance procedure service data unit (PSDU) to create a physical layer conformance procedure protocol data unit (PPDU) which is then transmitted. An AMPDU concatenates multiple AMPDU subframes that are created by adding delimiter and padding to each MPDU. All AMPDU subframes must be QoS data frames having the same TID and RA. The purpose of the delimiter is to allow enough time for the receiver to extract MPDUs while padding is done to make the AMPDU size a multiple of 4 bytes. The amount of delimiters added by a transmitter depends on the receiver's advertised minimum MPDU start spacing value in the HT capabilities IE, which defines the minimum time separation between consecutive MPDUs that the receiver can handle.

A transmitter can choose to perform AMSDU aggregation or AMPDU aggregation or combine both levels of aggregation depending on the use case and receiver's capabilities. For example, in scenarios where the receiver encounters high packet error rate (PER), AMPDU aggregation is better suited as the failed subframes can be identified and retransmitted. Table 2.4 describes the aggregation limits in Wi-Fi generations 4 and 5 including the range of possible values for the limit on each component.

**Table 2.4**
Aggregation Limits

| *Component* | *HT* | *VHT* |
|---|---|---|
| MSDU | 2304 bytes | 2304 bytes |
| MPDU | MAC header size + Maximum AMSDU length in HT capabilities IE | Maximum MPDU length in VHT capabilities IE |
| | MAC header size + 3839 or 7935 bytes | 3895 or 7991 or 11454 bytes |
| AMPDU | Maximum A-MPDU length exponent in HT capabilities IE | Maximum A-MPDU length exponent in VHT capabilities IE |
| | 8K to 64K bytes | 8K to 1M bytes |
| PPDU | 5.484 ms | 5.484 ms |

Before a transmitter can send a PPDU with AMPDU aggregation, it must setup a blockack (BA) session with the receiver by sending an add BA (ADDBA) request action frame and the recipient must respond with ADDBA response frame. Once a BA session is established, PPDUs with AMPDU can be sent and the recipient acknowledges them using either immediate or delayed BA frame instead of ACK. The BA frame allows to acknowledge multiple received MPDUs in one PPDU thereby reducing the overheads. The blockack policy of immediate or delayed BA is decided as part of the ADDBA request/response frame exchange. In the case of immediate BA, the recipient would send a BA frame in SIFS time upon receiving AMPDU. In the case of delayed BA, the transmitter has to send a blockack request (BAR) frame and the receiver would respond with BA in SIFS time. Both BAR and BA frames have a starting sequence number (SSN) field as shown in Figures 2.10, and 2.11 to indicate the smallest sequence number among the set of MPDUs to be acknowledged. The BA contains a 64-bit BA bitmap indicating FCS pass or fail for each of the MPDUs with sequence numbers from SSN to SSN+63. The BA session can be torn down by either transmitter or receiver by sending a delete BA (DELBA) action frame.

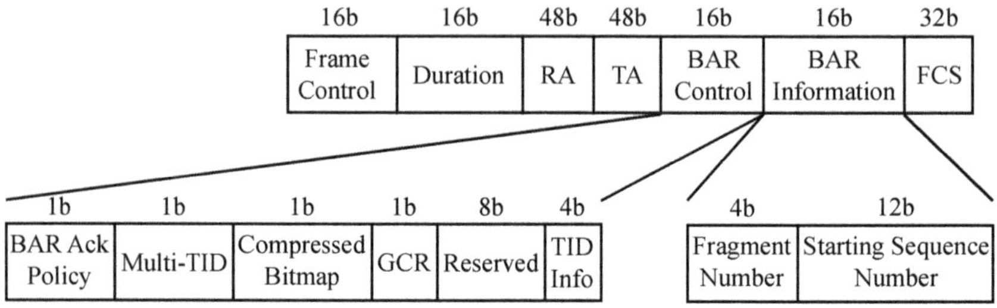


**Figure 2.10** BAR frame format.

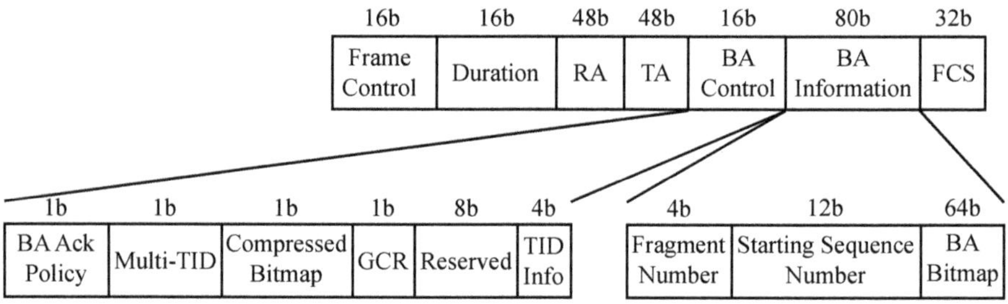


**Figure 2.11** BA frame format.

### 2.1.9 STA Power Save Methods

Wi-Fi standard provides various methods for battery-powered clients to save power. In the order of increasing power consumption, there are four power consumption states for any Wi-Fi STA: sleep, listen, receive, and transmit. Sleep is the lowest power consumption state during which STA doesn't transmit or listen to anything. In the listen state, the STA is actively searching for a Wi-Fi PHY preamble, and upon PHY preamble detection, the STA transitions to receive state during which it receives a Wi-Fi frame. The transmit state during which the STA transmits is the highest power consumption state.

#### 2.1.9.1 STA Power Save Notification

Wi-Fi standard gives ample flexibility for a STA to decide when to go to sleep. Before a STA goes to the sleep state, it should notify the AP that it is entering power save (PS) state by setting power management bit to 1 in the frame control field of MAC header. Typically a STA uses a null data frame for this notification although any data frame may be used. Upon receiving this notification, the AP starts buffering all intended frames for this STA instead of immediately transmitting them. After receiving ACK for the PS notification, the STA may enter the sleep state.

#### 2.1.9.2 Buffered Frame Notification by AP

The AP uses the traffic indication map (TIM) IE in the beacon frame to notify the STA of buffered frames. The TIM IE as shown in Figure 2.12 is present in every beacon frame and it contains a partial virtual bitmap field that includes AIDs of the STAs that have buffered traffic. Recall that the AID is a unique identifier assigned to each STA in the Association Response frame. In addition, every STA informs the AP how often it will wake up to listen for beacon in the listen interval field of the association request frame. The AP uses the listen interval field to determine the buffer size to allocate for STA in power save.

| 8b | 8b | 8b | 8b | 8b | 8-2008 b |
|---|---|---|---|---|---|
| Element ID | Length | DTIM Count | DTIM Period | Bitmap Control | Partial Virtual Bitmap |

**Figure 2.12** TIM IE format.

If there are any broadcast or multicast frames buffered at the AP, this is indicated by setting bit 0 of the bitmap control field of TIM IE, but this information is present only in certain beacons called delivery TIM (DTIM) beacons. DTIM beacons are sent periodically once every DTIM period, which is a multiple of beacon interval. The DTIM period is specified in the TIM IE along with a DTIM count field indicating how many beacons remain before the next DTIM beacon. The DTIM beacon has a DTIM count field value of 0. The AP transmits any buffered multicast and broadcast packets immediately after the DTIM beacon. All STAs are required to wake up every DTIM period and listen to the DTIM beacon, so this imposes a maximum limit on the sleep state duration.

### 2.1.9.3 Retrieving Buffered Frames

There are mainly three mechanisms for a STA to retrieve buffered frames from the AP: PS exit, PS poll and unscheduled automatic power save delivery (UAPSD) as illustrated in Figures 2.13, 2.14, and 2.16.

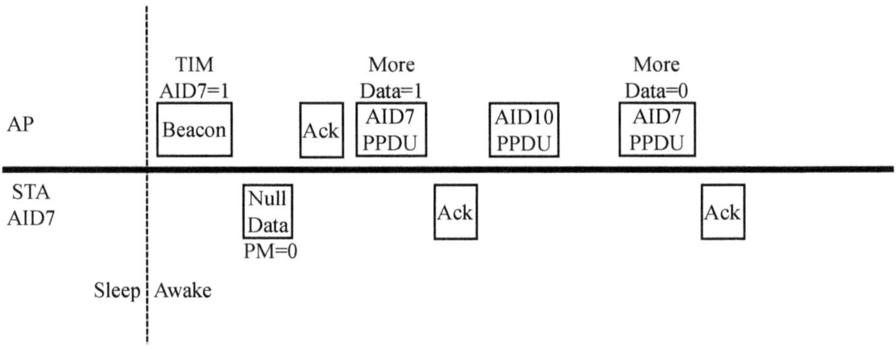


**Figure 2.13** Retrieving frames using PS exit.

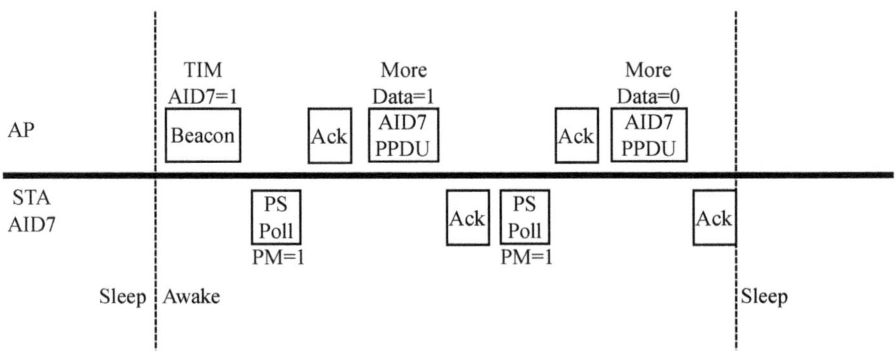


**Figure 2.14** Retrieving frames using PS poll.

In the PS exit mechanism, the STA sends a null data frame to AP with power management bit set to 0 indicating the STA has exited PS state and ready to receive frames. The AP then follows its own scheduling algorithm to send buffered frames to the STA, so the STA is unaware when exactly and for how long it will be serviced. When the AP sends the first buffered frame to the STA, it indicates whether there are more buffered frames using the more data bit in the frame control field of MAC header. But, since the STA doesn't know when the next buffered frame will be delivered, it remains in the listen state until it receives all buffered frames. AP sets the more data bit to 0 when there are no more buffered frames, after which the STA may enter PS state again. The PS exit mechanism is inefficient because the STA is in listen state for unknown time duration causing the fraction of time in PS state to be random depending on how many clients the AP is servicing.

The PS poll mechanism requires the STA to send a PS poll control frame to the AP which the AP acknowledges and immediately sends a PPDU with one buffered MSDU to the STA. The PS poll frame unlike other frames does not have a duration field and is instead replaced by STA AID as shown in Figure 2.15. AP uses the more data field to indicate buffered frame status. Unlike the PS exit mechanism, the power management bit is set to 1 in every frame from the STA. The advantage of PS poll mechanism is that the AP immediately services the PS client upon receiving PS poll but the disadvantage is that the client has to transmit a PS poll frame for every MSDU and AP cannot perform aggregation. PS poll may be a good scheme for very low traffic applications such as a temperature sensor but it is inefficient for moderate and high traffic applications like Voice over IP (VoIP), or Wi-Fi calling. The PS exit and PS poll mechanisms are sometimes referred as legacy PS mechanisms as they were part of the first 802.11 standard in 1997.

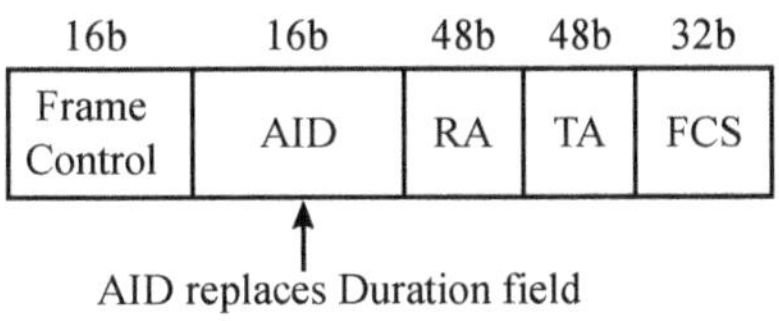


**Figure 2.15** PS poll frame format.

The UAPSD, also known as WMM PS, mechanism introduced as part of the 802.11e standard is applicable only for QoS data frames and requires support by both the AP and STA. An AP indicates UAPSD support in the WME QoS info IE of beacon while the STA indicates UAPSD support for each AC in the QoS info field of QoS capability IE in association request and reassociation request frames. An UAPSD-capable STA can choose to retrieve some or all buffered frames from the AP during an unscheduled service period, whose length is specified in the max SP length subfield of the QoS info field in the association request and reassociation request frames. The max SP length field is specified in terms of number of PPDUs and a STA can choose either 2 or 4 or 6 PPDUs or all buffered frames. The AP

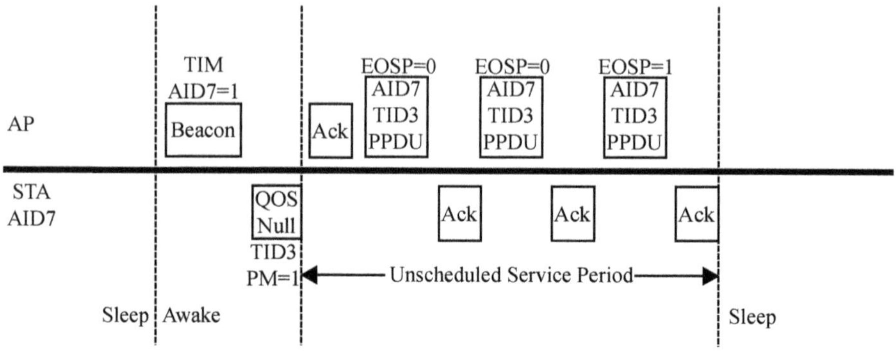


**Figure 2.16** Retrieving frames using UAPSD.

begins an unscheduled service period upon receiving a trigger frame from the STA. The trigger frame could be any QoS frame (either QoS data or QoS null) from the STA and the power management bit is always set to 1 similar to PS poll mechanism. During the unscheduled service period, the AP transmits QoS frames belonging to the particular TID specified in the trigger frame. The AP is permitted to send multiple PPDUs with aggregation as long as the duration is within the specified max SP length. The AP indicates end-of-service period (EOSP) to the STA using the EOSP bit in the QoS control field and uses the more data bit in frame control field to indicate whether there are buffered frames for any AC. Then, the STA can choose to send another trigger frame or go back to sleep. The UAPSD mechanism is suitable for Wi-Fi calling applications, where there is a predetermined number of frames to be transmitted periodically.

#### 2.1.9.4 Wireless Network Management Sleep Mode

Wireless network management sleep mode is a 802.11v feature that allows a STA to be in sleep state for an extended time without having to wake up to listen to beacons every listen interval or DTIM beacons while remaining associated to AP. An AP or STA can indicate wireless network management (WNM) sleep mode support by setting the WNM sleep mode field to 1 in the extended capabilities IE. The AP informs the maximum time period a client can remain associated without transmitting any frame to AP using the BSS max idle period IE in association response and reassociation response frames. The max idle period is specified in units of 1000 TUs. A STA may enter or exit WNM sleep mode by sending a WNM sleep mode request action frame to the AP. This action frame contains the requested WNM sleep interval, which indicates how often the STA will wake up to receive beacons. The AP confirms the WNM sleep mode request by sending a WNM sleep mode response action frame to the STA. WNM sleep mode is useful in applications such as Internet of Things (IoT), where it is important to extend battery life.

#### 2.1.9.5 TSF Synchronization

For power save methods to work effectively, all the STAs must be time synchronized to the AP so that the STA can receive beacons at the right time. If synchronization is lost, the STA could miss TBTT and it would have to be in the listen state for a longer period of time. To facilitate synchronization, all devices maintain a time synchronization function (TSF) timer which is a counter that increments every $\mu$s. The AP broadcasts the current value of its TSF timer in the 64-bit timestamp field of every beacon frame. Synchronization is achieved by requiring all connected STAs to update their TSF timer with the timestamp field value upon receiving a beacon frame. The need for synchronization is one of the main reasons why STAs have to wake up periodically to listen to beacons.

#### 2.1.9.6 Spatial Multiplexing Power Save

Spatial multiplexing power save (SMPS) or multiple input multiple output (MIMO) power save is a power save mechanism for 802.11n STAs supporting multiple spatial streams to save on listen state power by using only one receive chain. The PHY layer details on multiple spatial stream operation are provided in Section 2.2.4.3. There are two modes of SMPS: static SMPS and dynamic SMPS. In static SMPS mode, the STA enables only one receive chain and the AP transmits frames to this STA at one spatial stream data rates. In the dynamic SMPS mode, the STA listens on one active receive chain but switches to all receive chain mode upon receiving a frame destined to it. The STA goes back to single chain mode if there are no more frames destined to it. An AP can use an RTS-CTS frame exchange or any frame at a single spatial stream data rate to make the STA enable all receive chains. A STA indicates support for SMPS and the mode of SMPS used in the SMPS subfield of HT capabilities IE. A STA can change the SMPS mode after association by sending a SMPS action frame to the AP.

### 2.1.10 Multicast Traffic

While broadcast traffic is intended for all recipients, multicast traffic is intended for a group of interested recipients. Multicast traffic is useful in applications such as audio or video streaming and helps reduce the traffic on the medium. There is a bit in both IP address and MAC address to indicate whether a frame is multicast frame or unicast frame. The IP address range 224.0.0.0 to 239.255.255.255 is set aside for multicast groups and there is a method to map the multicast IP address to the multicast MAC address. There are 23 bits of MAC address space available to define $2^{23}$ unique multicast groups. Recipients can subscribe to a multicast group using the internet group management protocol (IGMP) and APs perform IGMP snooping to inspect IGMP frames and build a multicast table that map devices to multicast groups.

Sending multicast frames over Wi-Fi has several challenges and trade-offs. Multicast frames are not acknowledged, so the AP sends all multicast frames at the lowest data rate supported by all clients. Several AP implementations permit the user to configure the PHY rate for multicast frame. Therefore, multicast traffic tends to consume a significant amount of airtime especially for applications such as high definition (HD) video streaming. Since Wi-Fi has no error recovery mechanism for multicast frames, applications have to rely on higher layer error recovery and retransmission mechanisms. Multicast frames can only be sent at DTIM interval and all STAs have to wake up to receive it even if some STAs have not subscribed to any multicast group. If the DTIM interval is too high, then latency for multicast traffic increases. With low DTIM interval the battery life of STAs is reduced, and this is a trade-off. Most enterprise APs support a feature to convert multicast to unicast traffic as this will be more efficient on the air, especially when the number of clients in a multicast group is low.

### 2.1.11 STA Roaming

To provide Wi-Fi coverage in large areas such as offices, and large or multistory homes, multiple APs advertising the same SSID are deployed. Multiple APs with smaller coverage areas are also used to increase the capacity of the wireless network and such dense deployment is often done in stadiums, auditoriums, and other venues where large numbers of people gather. As a STA moves across the area, it should roam and associate to the closest possible AP to maintain a robust connection. The roaming algorithm that determines when to roam and which AP to connect is proprietary and different clients have subtle variations in the behavior. Typically, a

STA considers roaming when the RSSI of frames received from the AP falls below a certain threshold or if it fails to receive multiple consecutive beacons. At that point, STA starts to perform active and passive scanning to discover neighboring APs advertising the same SSID. If the STA finds a different AP with stronger RSSI, it may decide to roam. The medium utilization (percentage of time the medium is occupied) in a channel could also influence the STA's decision. After identifying a target AP to connect, the STA disconnects from the current AP, and associates and authenticates with the target AP on the new channel. The STA needs to complete roaming before it moves out of the range of the current AP; otherwise, it could result in a disconnect or disruption of service.

#### 2.1.11.1 Neighbor Report

To help the STA scan and identify a suitable target AP to roam, the IEEE 802.11k standard introduced radio resource management (RRM). Support for RRM is indicated using the RRM capability IE in beacon and association frames. If the AP and STA support RRM, a STA can request the AP to send a neighbor report using an action frame. The AP responds with a neighbor report containing a list of neighbor APs that advertise same SSID, along with their BSSID and channel information. The APs on the same network are aware of their neighbors as they are connected to each other via Ethernet or alternatively APs can discover neighbors using periodic off-channel scanning. The STA could then perform a much quicker scan limited to channels listed in the neighbor report, saving both power and time. RRM therefore helps in reducing roaming time and the amount of active scan traffic on all channels.

#### 2.1.11.2 BSS Transition Management

Sometimes APs are in a better position to recommend a target AP for a STA to roam. Even though two APs may be providing a healthy signal strength, the capacity available in the two APs could differ significantly if one of the APs is already serving a large number of connected clients. BSS transition management (BTM) was introduced in the IEEE 802.11v standard to address such scenarios, where an AP can influence STA's roaming behavior. BTM support is advertised in the extended capabilities IE of beacon, probe, and association frames. When BTM is supported on both AP and STA, the AP can send a BTM request to a STA either unsolicited or upon receiving a BTM query frame from the STA. The BTM request frame contains a recommended list of target APs to roam, along with their channel and BSSID information. However, it is up to the STA to accept or reject the

BTM request. Additionally, the AP can set the disassociation imminent bit to 1 and populate the disassociation timer field in the BTM request frame to inform the client that it will be disconnected from the AP after the time indicated in disassociation timer field. This option can influence clients to roam to the AP recommended in BTM request. A BTM capable STA looking to roam can reduce scanning time by sending a BTM query frame to solicit a BTM request frame from the AP and then roam to the recommended AP. BTM is useful to load-balance clients across multiple neighboring APs in a controller-managed enterprise network and across dual radios in a dual-band concurrent AP[1].

### 2.1.12 AP Channel Switching

An AP may sometimes want to change channel due to interference from cochannel APs or non-Wi-Fi devices (Bluetooth, microwave, etc). In some cases, APs are required by regulatory rules to change channels upon detection of a licensed incumbent device (for example radar in UNII-2 band). When an AP changes channel, the connected clients experience a beacon miss followed by disconnect event and would start scanning again. To avoid such disruption, the IEEE 802.11h standard introduced a channel switch announcement (CSA) IE in beacons and probe response frames to enable an AP to inform associated clients of an imminent channel change along with details of the new channel and the timing of channel change. This enables connected clients to seamlessly follow an AP to the new channel while maintaining the associated state. The CSA IE has a channel switch count field that enables AP to indicate the number of TBTTs pending for the channel change to occur. The AP sets the channel switch count field large enough to allow all PS STAs to receive at least one beacon containing CSA IE. Also, an AP can notify all connected STAs to stop further transmissions on the channel by setting the channel switch mode field to 1 in the CSA IE. The CSA IE can also be sent between beacons using a CSA action frame. Most clients and APs support the CSA mechanism of channel change.

## 2.2 PHY LAYER OVERVIEW

The PHY layer has a transmitter and receiver module. The transmitter module maps the PSDU from MAC layer to a RF waveform that is transmitted using an antenna

1 A dual-band concurrent AP is an AP with multiple radios capable of providing Wi-Fi service in two frequency bands concurrently.

as an electromagnetic wave. The receiver module essentially performs the reverse operation of transmitter module; that is, it decodes the received RF waveform, extracts the PSDU, and provides it to the MAC layer. All Wi-Fi standards specify the PHY transmitter in detail while only certain guidelines are given for receiver implementation. The PHY layer transmitter creates a PPDU by appending the PLCP header and PLCP preamble to the PSDU provided by the MAC layer. The PLCP preamble present at the beginning of every PPDU is a known waveform with special self-correlation properties that enable receivers to detect the beginning of a Wi-Fi frame and perform automatic gain control (AGC) to size the incoming signal and compensate frequency offset. The PLCP header contains information required to decode the rest of the PPDU such as data rate, bandwidth (BW), PPDU length, and so on. The PPDU is converted from bits to a RF waveform using a two-step process process called coding and modulation. Coding is the process of adding redundant bits to the data bits so that decoding errors at the receiver can be detected and corrected. The amount of errors that can be corrected increases with decreasing coding rate, which is defined as the ratio of data bits to coded bits. Although coding reduces the data rate, it is essential for a reliable communication link. Following the coding step, the modulation step maps a block of $n$ bits every symbol duration $T_s$ seconds to one of $2^n$ combinations of (amplitude, phase) of a sinusoidal waveform at desired RF. This mapping is pictorially represented as a two-dimensional plot called the constellation diagram, in which every constellation point has the amplitude represented by distance from origin and phase represented by angle made with the X axis. Figure 2.17 shows a sample constellation diagram for quadrature phase shift keying (QPSK) modulation that maps 2 bits to four different phases. The modulated waveform that is transmitted on the air occupies a certain frequency range, which is defined as BW. One of the important metrics for PHY layer is the spectral efficiency, which is defined as the ratio of data rate to BW. The coding and modulation used, format of PPDU, PLCP preamble, and PLCP header depend on the Wi-Fi generation.

### 2.2.1 Wi-Fi Generation 1 (802.11b)

Wi-Fi generation 1 operates in the 2.4-GHz band and is based on the first 802.11 standard and 802.11b amendment released in 1997 and 1999, respectively. It uses DSSS technology and offers four different data rates of 1 Mbps, 2 Mbps, 5.5 Mbps, and 11 Mbps. There is no error correction code employed and the occupied transmission BW is 22 MHz. The PLCP header is 48 bits long and there are two

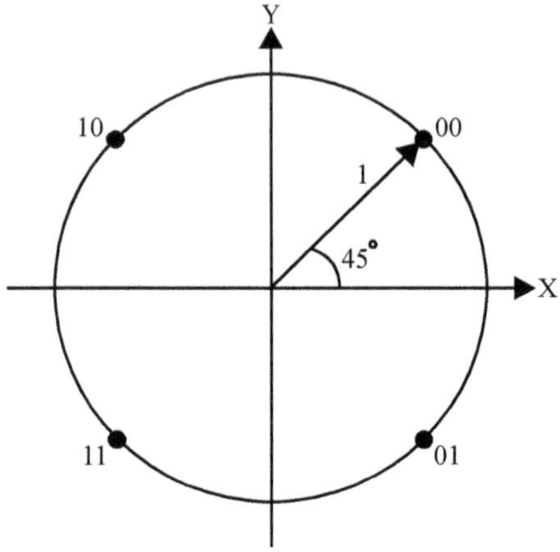


**Figure 2.17** QPSK constellation diagram.

variants available for the PLCP preamble: long preamble and short preamble as shown in Figure 2.18.

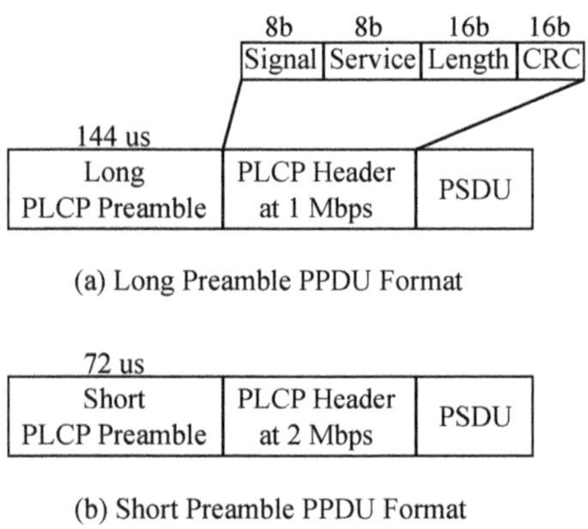


**Figure 2.18** Wi-Fi 1 (802.11b) PPDU format.

In the long preamble variant, the PLCP preamble is 144 $\mu$s long and the PLCP header is sent at 1-Mbps data rate. In the short preamble variant, the PLCP preamble is 72 $\mu$s long and the PLCP header is sent at 2-Mbps data rate. The 48-bit PLCP header has a 8-bit signal field to indicate the data rate for PSDU, 16-bit length field to specify PSDU duration in $\mu$s, and a 16-bit CRC field for integrity check covering the PLCP header. The data rate for ACK and CTS control frames is mandated to be 1 Mbps.

### 2.2.2 Wi-Fi Generation 2 (802.11a)

Wi-Fi generation 2 is based on the 802.11a standard released in the same year as 802.11b, but it took a few more years for consumer devices to broadly adopt it due to increased implementation cost. The 802.11a PHY is based on OFDM technology, which is a multicarrier modulation scheme. In multicarrier modulation, the transmission BW is divided into narrow subchannels and the information bit stream is divided into an equal number of parallel bit streams. The center frequency of each subchannel is called subcarrier. Each bit stream modulates one of the subcarriers and the modulated outputs across all bit streams are summed up. OFDM is a specific form of multicarrier modulation, where the subchannels are half overlapping but yet orthogonal resulting in high spectral efficiency. In order to combat inter symbol interference (ISI) due to multipath, a cyclic prefix (CP) or guard interval (GI) of duration greater than the wireless channel's delay spread is added as a prefix to the transmitted waveform. Channel delay spread is defined as the difference between the earliest arriving multipath component (example line of sight) and the latest arriving multipath component. Although GI is an overhead, it significantly simplifies the demodulation process at the receiver. Multicarrier modulation and demodulation in OFDM can be implemented using efficient inverse fast Fourier transform (IFFT) and fast Fourier transform (FFT) operations.

The 802.11a PHY employs OFDM with 64 subcarriers, 0.8-$\mu$s GI, modulation options from binary phase shift keying (BPSK) to 64 quadrature amplitude modulation (QAM), and has a transmission BW of 20 MHz. The subcarrier frequency spacing is 20 MHz/64 = 312.5 kHz and the OFDM symbol duration is 1/312.5 kHz = 3.2 $\mu$s. Out of the 64 subcarriers, 48 subcarriers are used for data transmission, 12 subcarriers are left unused, and 4 subcarriers are assigned as pilots to enable the receiver to track wireless channel variations with time. A convolutional code is used for error correction with coding rate options of 1/2, 2/3, and 3/4. There are eight supported data rates as described in Table 2.5. The data rate for ACK and CTS frames are determined based on the modulation of received PPDU and RTS frames as described in Table 2.6.

The 802.11a PPDU format is shown in Figure 2.19. The PLCP preamble consists of 8-$\mu$s long legacy short training field (L-STF) for frame detection, AGC, coarse frequency offset correction and 8-$\mu$s long legacy long training field (L-LTF)

**Table 2.5**

Wi-Fi 2 to Wi-Fi 5 Data Rates in Mbps for 1 Spatial Stream

| *Modulation* | *Coding Rate* | *GI ($\mu s$)* | *Legacy* | *HT MCS* | *VHT MCS* | Bandwidth (MHz) HT/VHT 20 | 40 | VHT 80 | 160 |
|---|---|---|---|---|---|---|---|---|---|
| BPSK | 1/2 | 0.8 | 6 | 0 | 0 | 6.5 | 13.5 | 29.3 | 58.5 |
| BPSK | 1/2 | 0.4 | NA | 0 | 0 | 7.2 | 15 | 32.5 | 65 |
| BPSK | 3/4 | 0.8 | 9 | NA | NA | | | | |
| QPSK | 1/2 | 0.8 | 12 | 1 | 1 | 13 | 27 | 58.5 | 117 |
| QPSK | 1/2 | 0.4 | NA | 1 | 1 | 14.4 | 30 | 65 | 130 |
| QPSK | 3/4 | 0.8 | 18 | 2 | 2 | 19.5 | 40.5 | 87.8 | 175.5 |
| QPSK | 3/4 | 0.4 | NA | 2 | 2 | 21.7 | 45 | 97.5 | 195 |
| 16-QAM | 1/2 | 0.8 | 24 | 3 | 3 | 26 | 54 | 117 | 234 |
| 16-QAM | 1/2 | 0.4 | NA | 3 | 3 | 28.9 | 60 | 130 | 260 |
| 16-QAM | 3/4 | 0.8 | 36 | 4 | 4 | 39 | 81 | 175.5 | 351 |
| 16-QAM | 3/4 | 0.4 | NA | 4 | 4 | 43.3 | 90 | 195 | 390 |
| 64-QAM | 2/3 | 0.8 | 48 | 5 | 5 | 52 | 108 | 234 | 468 |
| 64-QAM | 2/3 | 0.4 | NA | 5 | 5 | 57.8 | 120 | 260 | 520 |
| 64-QAM | 3/4 | 0.8 | 54 | 6 | 6 | 58 | 121.5 | 263.3 | 526.5 |
| 64-QAM | 3/4 | 0.4 | | 6 | 6 | 65 | 135 | 292.5 | 585 |
| 64-QAM | 5/6 | 0.8 | | 7 | 7 | 65 | 135 | 292.5 | 585 |
| 64-QAM | 5/6 | 0.4 | | 7 | 7 | 72.2 | 150 | 325 | 650 |
| 256-QAM | 3/4 | 0.8 | | NA | 8 | 78 | 162 | 351 | 702 |
| 256-QAM | 3/4 | 0.4 | | NA | 8 | 86.7 | 180 | 390 | 780 |
| 256-QAM | 5/6 | 0.8 | | NA | 9 | NA | 180 | 390 | 780 |
| 256-QAM | 5/6 | 0.4 | | NA | 9 | NA | 200 | 433.3 | 866.7 |

**Table 2.6**

Data Rate for ACK and CTS frames

| *Modulation of received PPDU/RTS frame* | *ACK/CTS Data Rate (Mbps)* |
|---|---|
| BPSK | 6 |
| QPSK | 12 |
| 16 QAM or higher | 24 |

for fine frequency offset correction, channel estimation at the receiver. The PLCP header is a 24-bit legacy signal (L-SIG) field sent at 6-Mbps data rate and contains information on the PSDU data rate and PSDU length in bytes along with an integrity check.

| 8 us | 8 us | 4 us | |
|---|---|---|---|
| L-STF | L-LTF | L-SIG | PSDU |

**Figure 2.19** Wi-Fi 2 (802.11a) PPDU format.

### 2.2.3 Wi-Fi Generation 3 (802.11g)

Although as a technology, OFDM in Wi-Fi 2 proved to be superior than DSSS in Wi-Fi 1, consumer adoption turned out to be very slow because Wi-Fi 2 required adding a 5-GHz band radio, which was expensive at that time. To address the immediate need for lower cost and higher data rates in the 2.4-GHz band, the 802.11g standard was introduced in 2003. Wi-Fi generation 3 based on 802.11g added some minor changes for 802.11b compatibility to make the 802.11a OFDM PHY work in 2.4-GHz band. All Wi-Fi 3 devices are required to support Wi-Fi 1 and employ 1 Mbps data rate for all broadcast frames and management frames. Wi-Fi 3 achieved backward compatibility by adding a protection mode in which all OFDM transmissions are protected with a RTS-CTS frame exchange at 1-Mbps data rate, so that Wi-Fi 1 devices will not interfere. A Wi-Fi 3 AP indicates to all STAs that they must employ protection mode by setting the protection mode subfield to 1 in ERP IE of beacon frame. Moreover, an AP indicates all STAs to use 20-$\mu$s slot time by setting short slot Time subfield to 0 in the capability info IE of beacons. A Wi-Fi 3 AP advertises protection mode and disallows a short slot time upon detection of a Wi-Fi 1 AP in its vicinity or if a Wi-Fi 1 STA connects to it. In the absence of Wi-Fi 1 devices, Wi-Fi 3 APs may use 9-$\mu$s slot time (short slot time=1) to take advantage of higher throughput. Since SIFS time is 16 $\mu$s in Wi-Fi 2 while it is only 10 $\mu$s in the 2.4-GHz band, Wi-Fi 3 devices are required to add a 6-$\mu$s silence period to the end of all OFDM transmissions to allow enough time for receivers to process the frame and generate response frame. Due to its reuse of Wi-Fi 2 to a large part with only minor changes, Wi-Fi 3 devices were quick to market and gained phenomenal consumer traction.

### 2.2.4 Wi-Fi Generation 4 (802.11n)

In 2009, Wi-Fi 4 based on 802.11n was introduced and it supported operation in both the 2.4-GHz and 5-GHz bands. Wi-Fi 4 significantly increased the spectral efficiency using a new feature called spatial multiplexing based on MIMO OFDM technology. It is well known [2] that the spectral efficiency of any communication system scales logarithmically with the signal-to-noise ratio (SNR) at the receiver. An exponential increase in SNR (or equivalently transmit power) therefore provides only a linear increase in spectral efficiency. MIMO is an alternate technique to increase the spectral efficiency by transmitting and receiving multiple data streams concurrently using multiple antennas. In 1998, researchers at Bell Laboratories demonstrated [3] a practical MIMO communication system whose spectral efficiency scales linearly with the minimum of the number of transmit and receive antennas. Wi-Fi 4 combines MIMO with OFDM to enable high spectral efficiency along with multipath resilience. Wi-Fi 4 also added other PHY layer features such as 40 MHz transmission BW, short GI, and low density parity check (LDPC) to increase the data rate and error correction performance. The 802.11n PHY layer is called the HT PHY layer, so any frame transmitted using Wi-Fi 4 data rate is also referred as a HT frame.

#### 2.2.4.1 Error Correction Code

Wi-Fi 4 supports an optional error correction coding scheme called LDPC while support for convolutional code is mandatory. LDPC code provides incrementally better error rate performance compared to convolutional code. The supported coding rates are 1/2, 2/3, 3/4, and 5/6. LDPC reception capability is advertised in the HT capabilities IE.

#### 2.2.4.2 Short GI

Wi-Fi 4 offers two choices for the OFDM GI: 0.8 $\mu$s and an optional 0.4-$\mu$s GI (referred as half GI or short GI) for higher data rate in certain multipath environments having low delay spread. However, the standard does not describe the logic to decide GI, and this is left open to implementation. Short GI reception capability is advertised in the HT capabilities IE.

#### 2.2.4.3 Spatial Multiplexing

A Wi-Fi 4 device with $N_{Tx}$ transmit antennas and $N_{Rx}$ receive antennas (referred as $N_{Tx} \times N_{Rx}$ MIMO capable) can use spatial multiplexing [3] to transmit up to $N_{Tx}$ independent data streams and receive up to $N_{Rx}$ data streams. The maximum number of spatial streams (NSS) supported on transmit and receive are advertised by the AP and STA in the HT capabilities IE. Wi-Fi 4 limits the maximum NSS to 4. The NSS in each transmission must be less than or equal to the minimum of the transmitting device's $N_{Tx}$ and the intended receiving device's stream capability. The block diagram of a spatial multiplexing transmitter is shown in Figure 2.20. The data bits after error correction coding are split into NSS parallel coded data streams. Each coded data stream is then interleaved and mapped to constellation points (BPSK, QPSK, 16-QAM or 64-QAM). The resulting NSS constellation streams are expanded to $N_{Tx}$ streams using a matrix operation known as spatial expansion. A cyclic shift diversity (CSD) operation is applied on each of these $N_{Tx}$ streams after they are modulated using OFDM. The CSD operation essentially adds an antenna dependent time delay to the OFDM symbols. The CSD technique is especially useful when NSS $= 1$ and $N_{Tx} > 1$, wherein the same information is repeated on $N_{Tx}$ antennas leading to constructive or destructive combining of subcarriers at some receiver locations. Since the time delay introduced by CSD produces varying phase shift across subcarriers, it avoids the scenario of all subcarriers simultaneously experiencing a null at the receiver. To completely avoid nulling at any subcarrier for the NSS $= 1$, $N_{Tx} > 1$ scenario, Wi-Fi 4 supports an optional feature called space time block code (STBC), but this new feature was not broadly supported by many devices due to higher implementation complexity.

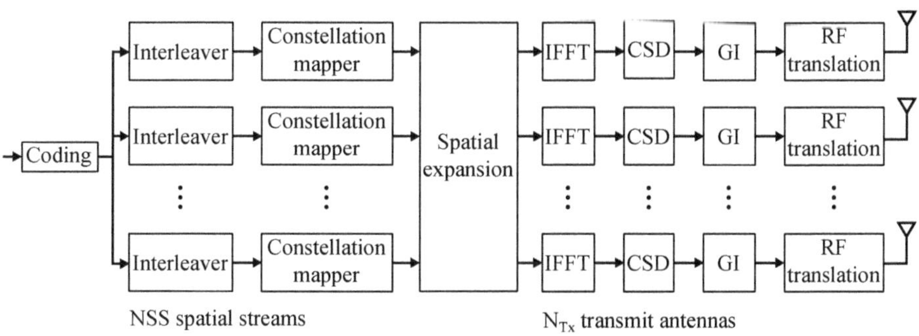


**Figure 2.20** Block diagram of a spatial multiplexing transmitter.

Next we describe how a MIMO receiver separates the multiple spatial streams transmitted by the MIMO transmitter. The 802.11n standard does not describe a spatial multiplexing receiver and leaves it as an implementation choice. The underlying principle that enables a receiver to separate the multiple spatial streams is that the receiver has multiple antennas and each antenna receives a different combination of the spatial streams as a result of the spatial separation between the antennas and the multipath nature of RF propagation. If the antennas are separated by more than half the wavelength corresponding to the channel frequency, each antenna is likely to receive statistically independent combinations of the multiple spatial streams. Assuming high SNR, the additive noise component at each antenna can be ignored for simplicity and the spatial stream separation problem at the receiver becomes equivalent to solving for NSS unknown variables using $N_{Rx}$ equations. It is well known that the solution for such a system of equations is possible when the number of equations ($N_{Rx}$) exceeds or equals the number of unknown variables (NSS). This explains why NSS has to be less than or equal to the receiving device's $N_{Rx}$. Figure 2.21 illustrates how a spatial multiplexing receiver can separate multiple spatial streams with an example.

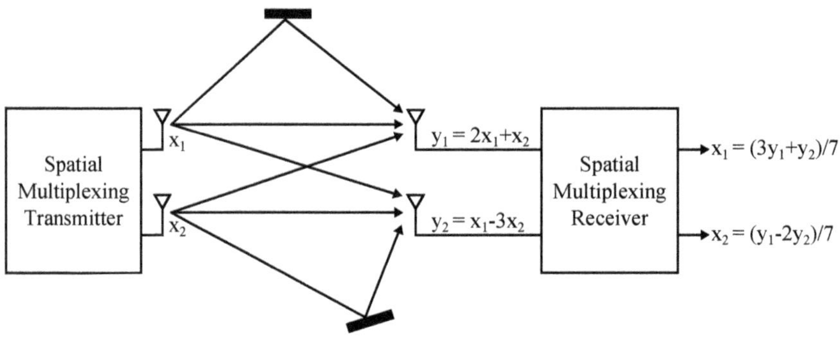


**Figure 2.21** Sample illustration of spatial multiplexing receiver.

### 2.2.4.4 HT PPDU Format

The HT PPDU format is shown in Figure 2.22. The first 20 $\mu$s of Wi-Fi 4 PLCP preamble is the Wi-Fi 2 PLCP preamble and header, so all OFDM legacy devices can decode the L-SIG field and backoff from transmission. The data rate and length fields in L-SIG are populated such that the calculated legacy PPDU duration covers the PSDU portion and the Wi-Fi 4 PLCP preamble. The remainder of the PPDU can be understood only by Wi-Fi 4 or later generation devices. The 8-$\mu$s long HT-SIG field is BPSK modulated and contains information required to decode the PSDU portion such as modulation and coding scheme (MCS), GI, BW, STBC, LDPC, PSDU length, along with 8-bit CRC protection. Following HT-SIG, the 4-$\mu$s long HT-STF field is for finer AGC. The HT-LTF portion is used for MIMO channel estimation and the number of HT-LTFs in a HT PPDU is equal to or greater than the NSS.

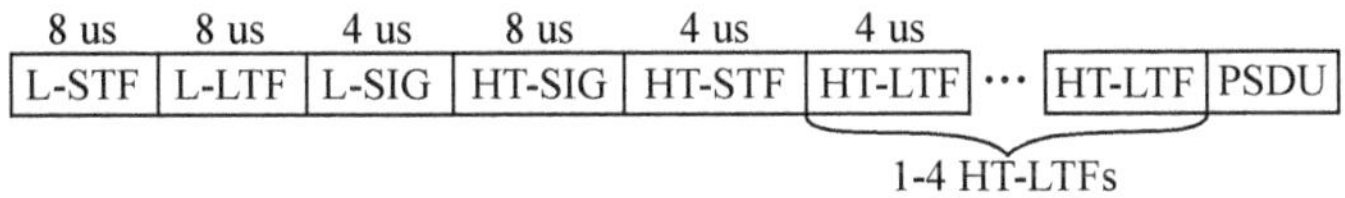


**Figure 2.22** Wi-Fi 4 (802.11n) HT PPDU format.

### 2.2.4.5 40 MHz Transmission Bandwidth

Wi-Fi 4 supports two transmission BW modes: 20 MHz and 40 MHz. In the 20-MHz transmission BW mode, each OFDM symbol has 64 subcarriers out of which 4 are pilot subcarriers and 52 are data subcarriers as compared to 48 data subcarriers in Wi-Fi 3. In the 40-MHz transmission mode, each OFDM symbol has 128 subcarriers, out of which there are 108 data subcarriers and 6 pilot subcarriers. A 40-MHz channel is created by bonding two contiguous 20-MHz channels.

One of these two 20-MHz subchannels is assigned as the primary channel or control channel and the other is referred to as the secondary or extension channel. All broadcast and management frames such as beacons and probes are sent only on the primary channel. The location of the secondary channel relative to the primary channel is specified as a positive or negative offset in the HT operation IE of beacon frame. To ensure backward compatibility, the PLCP preamble portion of every 40-MHz BW frame is transmitted on both 20-MHz subchannels, so that devices with only 20-MHz capability can coexist. Similarly, RTS-CTS protection and ACK frames for a 40-MHz bandwidth frame are duplicated on both the 20-MHz subchannels. This PPDU format involving duplication across of legacy frame across all 20-MHz subchannels is called a non-HT duplicate PPDU format. Backward compatibility with Wi-Fi 1 devices is maintained using similar mechanisms as in Wi-Fi 3.

Before any 40-MHz transmission, EDCA should declare the medium as idle on both 20-MHz subchannels. The supported Wi-Fi 4 data rates for one spatial stream are summarized in Table 2.5. To simplify describing more than 100 different data rates, Wi-Fi 4 proposed combining the modulation, coding rate, and NSS information into one number called HT MCS. For example, HT MCS 0-7 are for one spatial stream, HT MCS 8-15 are for two spatial stream, and so on.

### 2.2.5 Wi-Fi Generation 5 (802.11ac)

The fifth generation of Wi-Fi was introduced in 2013 based on the 802.11ac standard and supports only the 5-GHz band. Wi-Fi 5 is backward compatible with Wi-Fi 4 and Wi-Fi 2 devices. The Wi-Fi 5 PHY layer is called VHT PHY. Wi-Fi 5 primarily expanded on the ideas of MIMO and BW increase in Wi-Fi 4 to significantly increase the data rate. It added support for 80-MHz and 160-MHz transmission BW while also increasing the maximum NSS to 8. Optional support for 256-QAM modulation was also added. The OFDM subcarrier allocation, GI options, and coding options in Wi-Fi 5 are the same as in Wi-Fi 4. For the newly added 80-MHz BW transmission mode, each OFDM symbol has 256 subcarriers out of which 234 are data and 8 are pilots. The 160-MHz BW mode subcarrier allocation is exactly double that of 80-MHz with 468 data subcarriers and 16 pilot subcarriers. The definition of MCS in Wi-Fi 5 was modified to represent only modulation and coding rate, therefore the NSS has to be specified separately. The supported Wi-Fi 5 data rates for one spatial stream are summarized in Table 2.5. The supported NSS, BW, BW, and capabilities such as LDPC, short GI, transmit beamforming (TxBf), and multiuser MIMO (MU-MIMO) are advertised using the VHT capabilities IE. The new technologies introduced in Wi-Fi 5 are TxBf and downlink (DL) MU-MIMO. Although technically, 802.11n was the first standard to introduce TxBf feature, it was unsuccessful due to multiple optional variants leading to interoperability issues between AP and client vendors. These mistakes were corrected in Wi-Fi 5 by supporting only one variant of TxBf and ensuring interoperability through WFA certification.

#### 2.2.5.1 VHT PPDU format

Figure 2.23 shows the VHT PPDU format. Similar to Wi-Fi 4, the legacy OFDM PLCP preamble is present in the beginning for backward compatibility. The 8-$\mu$s long VHT-SIG-A field is sent at a 6-Mbps data rate and contains information on PSDU such as NSS, MCS, BW, short GI, LDPC, and beamforming indication along with 8-bit CRC protection. It also has a partial AID to identify the intended recipient in DL MU-MIMO transmission. The 4-$\mu$s long VHT-STF is meant for AGC updates. The VHT-LTFs are for MIMO channel estimation and the number of VHT-LTFs present is greater than or equal to NSS. The VHT-SIG-B field specifies the PSDU length and in the case of MU-MIMO it specifies the MCS for all intended users. The legacy PLCP preamble and VHT-SIG-A fields are duplicated in each 20-MHz subchannel for transmission bandwidths greater than 20 MHz.

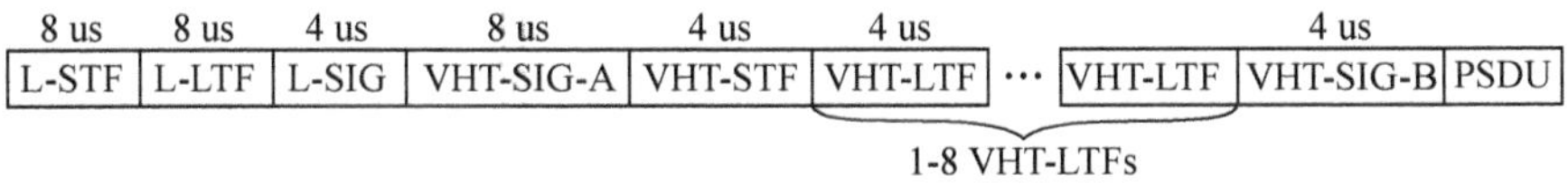


**Figure 2.23** Wi-Fi 5 (802.11ac) VHT PPDU format.

### 2.2.5.2 Transmit Beamforming

TxBf is a transmission technique applicable only when $N_{Tx} > 1$ and NSS is less than $N_{Tx}$. The device using TxBf technique for transmission of a frame is called the beamformer, while the device receiving the beamformed frame is called the beamformee. In this technique, the beamformer adjusts the amplitude and phase of the waveforms at each of the $N_{Tx}$ antennas in such a way that each spatial stream experiences coherent combining at every subcarrier at the beamformee location. One way to conceptually understand TxBf is that it performs amplitude and phase adjustments at multiple antennas to effectively create an equivalent directional antenna that focuses the radiated energy towards the beamformee.

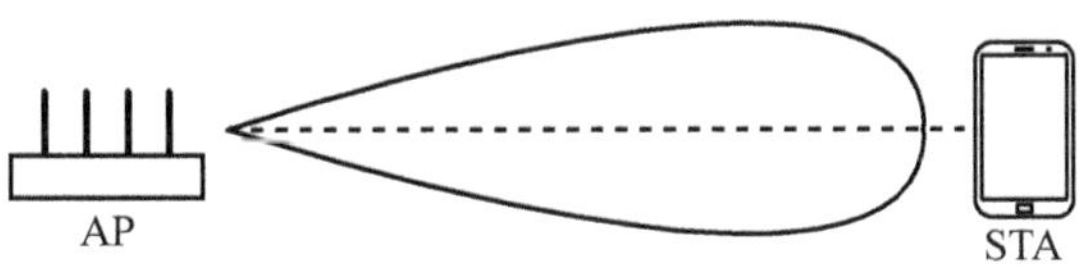


**Figure 2.24** Transmit beamforming.

To achieve coherent combining at the beamformee, the beamformer has to know the amplitude and phase changes caused by the wireless medium between each of the beamformer antennas and the beamformee antennas. The beamformer obtains this information by a periodic channel sounding process as illustrated in Figure 2.25. The beamformer initiates channel sounding by transmitting the null data packet announcement (NDPA) frame, which specifies the beamformee by AID in the case of a STA and 0 in the case of an AP. Following the NDPA frame the beamformer transmits a null data packet (NDP) frame in SIFS time interval. The VHT-LTFs in the NDP frame are used by the beamformee to estimate the wireless channel and calculate a compressed beamforming (CBF) report that contains the average SNR for each spatial stream quantized to 8 bits and a compressed version of the estimated wireless channel matrix represented as a list of angles. The beamformee then transmits this CBF report to the beamformer using a compressed beamforming action frame. After receiving the CBF report, the beamformer derives a steering matrix that is used for spatial expansion (see Figure 2.20) in subsequent beamformed frame transmissions. TxBf is not applied to the portion prior to VHT-SIG-B field in the VHT PPDU format, so that all devices in the vicinity can hear the PLCP preamble for CSMA-CA.

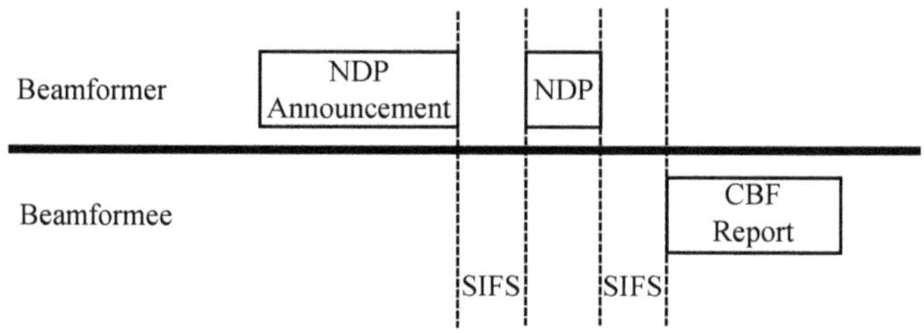


**Figure 2.25** TxBf channel sounding.

In theory, TxBf can provide maximum SNR gains of $10\cdot\log_{10}(\frac{N_r}{NSS})$ dB where, $N_r$ is the maximum number of beamformer antennas for which beamforming is supported by the beamformee. The effective performance gain in practice depends on (1) the quality of channel feedback that becomes worse at low SNR and (2) the airtime overhead for channel sounding that increases with the number of clients. The size of the CBF report in bits is given by (2.1).

$$\text{CBF report size in bits} = 8 \cdot N_c + N_a \cdot N_s \cdot (\text{Bit precision}) \quad (2.1)$$

where,

1. $N_c$ is the maximum NSS for which beamforming is intended.
2. $N_a$ is the number of angles in the CBF report, which is a function (see Table 2.7) of $N_c$ and the number of beamformer antennas $N_r$ for which beamforming is supported.
3. $N_s$ is the number of subcarriers for which channel feedback is provided, which in turn depends (see Table 2.8) on the BW and subcarrier grouping parameter $N_g$.

Subcarrier grouping is one way to reduce the CBF report size by averaging the information of $N_g$ adjacent subcarriers. VHT offers three choices for $N_g$; namely 1, 2, or 4 and two choices for the bit precision (3 or 5 bits). Although the beamformer has some control over the CBF report size by specifying bit precision and $N_g$ in the

**Table 2.7**

$N_a$ for Different $N_r$, $N_c$

| $N_r$ | $N_c$ | $N_a$ |
|---|---|---|
| 2 | 1 | 2 |
| 3 | 1 | 4 |
| 4 | 1 | 6 |
| 8 | 1 | 14 |
| 2 | 2 | 2 |
| 3 | 2 | 6 |
| 4 | 2 | 10 |
| 8 | 2 | 26 |
| 4 | 3 | 12 |
| 8 | 3 | 36 |

**Table 2.8**

$N_s$ for Different BW, $N_g$

| *BW (MHz)* | $N_g$ | $N_s$ |
|---|---|---|
| 20 | 1 | 52 |
| 20 | 2 | 30 |
| 20 | 4 | 16 |
| 40 | 1 | 108 |
| 40 | 2 | 58 |
| 40 | 4 | 30 |
| 80 | 1 | 234 |
| 80 | 2 | 122 |
| 80 | 4 | 62 |
| 160 | 1 | 468 |
| 160 | 2 | 244 |
| 160 | 4 | 124 |

NDPA frame, the actual airtime consumed by the CBF report also depends on the PHY rate used by the beamformee to send the CBF action frame.

#### 2.2.5.3 Downlink MU-MIMO

Typically APs have more antennas and can support higher NSS as compared to clients, which are constrained by power and physical dimensions. Such client side limitations often prevent an AP from realizing its maximum spatial multiplexing capabilities. The DL MU-MIMO feature introduced in Wi-Fi 5 enables an AP to transmit independent spatial streams to different users simultaneously. While TxBf steers all spatial streams towards the beamformee, DL MU-MIMO attempts to steer each spatial stream in different directions such that at each user location, all spatial streams except the respective user's stream will experience a null. This technique to mitigate inter-user interference is called null-steering.

Similar to TxBf, the AP (MU beamformer role) performs channel sounding and obtains a CBF report from each of the MU beamformee capable STAs. As shown in Figure 2.27, channel sounding for MU-MIMO is similar to TxBf sounding and starts with NDPA frame. But the NDPA frame in multiuser (MU) sounding has a broadcast RA field and it specifies the AIDs of more than one user which are supposed to prepare the channel feedback using the following NDP frame. The MU beamformer queries each STA to retrieve the channel feedback using

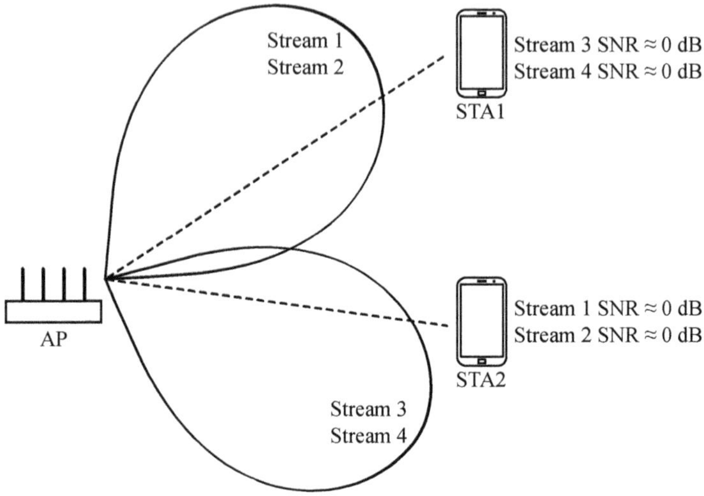


**Figure 2.26** Null-steering in downlink MU-MIMO.

a beamforming report poll frame. There are some subtle differences in content between MU CBF report and single user (SU) TxBf CBF report, so the requested feedback type is specified as MU in the NDPA frame. One such difference between MU and SU TxBf feedback is the higher bit precision used in MU CBF report, which can be either 6 or 8 bits. So, the size of MU CBF report is higher than that for SU TxBf CBF report.

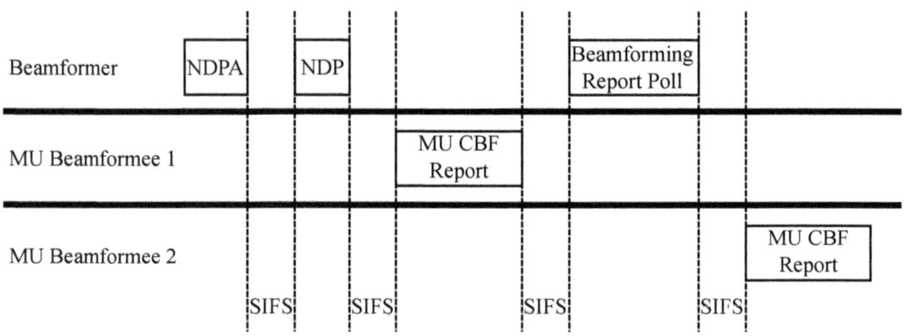


**Figure 2.27** MU-MIMO channel sounding.

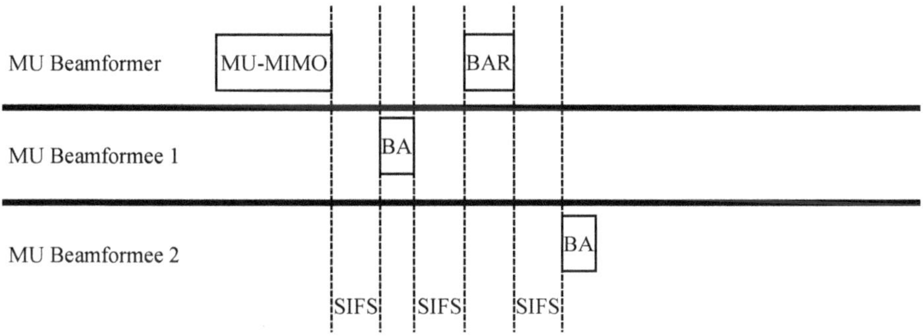


**Figure 2.28** Downlink MU-MIMO transmission.

Upon obtaining the channel feedback from a group of MU beamformees, the MU beamformer applies null-steering for subsequent MU-MIMO transmissions The AP then retrieves BA from each STA using a BAR frame as shown in Figure 2.28. In a MU-MIMO transmission, the total number of streams summed across all users has to be lower than the maximum NSS supported by the AP. Wi-Fi 5 limits the maximum number of users in a MU-MIMO transmission to 4. Despite the additional overheads, DL MU-MIMO is better than SU particularly for short-range and high traffic scenarios as it increases the spectral efficiency of the AP's transmissions and reduces time spent on RBO. However, one disadvantage is that DL MU-MIMO is sensitive to client mobility as it relies on precise channel feedback. To reap the best performance of DL MU-MIMO, an AP should adapt the frequency of MU sounding to match the STA mobility and employ an optimized MU grouping algorithm to decide the set of STAs in a DL MU-MIMO transmission, while also ensuring fairness in resource allocation.

## 2.3 REFERENCES

[1] "IEEE Standard for Information Technology–Telecommunications and Information Exchange between Systems - Local and Metropolitan Area Networks–Specific Requirements - Part 11: Wireless LAN Medium Access Control (MAC) and Physical Layer (PHY) Specifications," *IEEE Std 802.11-2020 (Revision of IEEE Std 802.11-2016)* (2021), pp. 1–4379, DOI: 10.1109/IEEESTD.2021.9363693.

[2] C.E. Shannon and W. Weaver, *The Mathematical Theory of Communication*, University of Illinois Press, 1949.

[3] P.W. Wolniansky et al., "V-BLAST: an architecture for realizing very high data rates over the rich-scattering wireless channel," *Proceedings of URSI International Symposium on Signals, Systems, and Electronics*, Oct. 1998.

# Chapter 3

# Wi-Fi 6

Wi-Fi 6 is the sixth generation of Wi-Fi introduced in 2019 based on the IEEE 802.11ax standard [1]. While the earlier generations of Wi-Fi focused on increasing peak data rates, Wi-Fi 6 is primarily about increasing efficiency in a network with multiple APs (physical and virtual), multiple clients, and varied traffic loads. The Wi-Fi 6 physical (PHY) is called high efficiency (HE) PHY and it strives to maximize every possible efficiency metric at the PHY layer while maintaining backward compatibility with previous Wi-Fi generations. Though improving efficiency is the primary objective, Wi-Fi 6 also increases the peak data rate by leveraging 1024-QAM modulation. Wi-Fi 6 addresses some long-standing issues, such as high airtime occupied by management traffic, low airtime efficiency in short data frames, and airtime wastage due to collisions and random backoff (RBO), commonly encountered in high density deployments. In the first four generations of Wi-Fi, sharing resources in the time dimension was the only way for multiple clients to share the medium. With MU-MIMO introduced in Wi-Fi 5, spatial streams became yet another dimension for multiple clients to share the medium, but this was limited only to downlink (DL) traffic. Up until Wi-Fi 5 [2], the access point (AP) has no control over the scheduling of uplink (UL) traffic and all clients have to contend for medium access. With orthogonal frequency division multiple access (OFDMA) technology and MU-MIMO technology, Wi-Fi 6 enables the AP to do resource allocation for downlink and uplink traffic [1, 3] in three dimensions; namely time, spatial streams, and frequency. OFDMA is a multiple access technique that splits the OFDM subcarriers into multiple groups called resource units (RUs) that can be assigned to different users. Since the subcarriers in OFDM are mutually orthogonal, each user can independently demodulate only their respective RUs. Wi-Fi 6 also introduces new power save mechanisms for AP to schedule station (STA) wake-up times, thereby reducing STA power consumption and collisions on the medium. Wi-Fi 6 introduces a new mode of multiple virtual access point (VAP) operation called enhanced multi-BSSID advertisement (EMA) to reduce probe response and beacon traffic in multiple VAP or multiple BSSID operation. Finally, to facilitate increased spatial reuse in high density AP deployments, Wi-Fi 6 introduces BSS coloring and spatial reuse. In summary, Wi-Fi 6 offers several new features with enough flexibility to address a wide range of use cases across varied deployments.

## 3.1 HE PPDU FORMATS

Wi-Fi 6 defines four types of PPDU formats as shown in Figure 3.1:

1. HE single user (SU) PPDU for SU;
2. HE ER SU PPDU for extended range SU;
3. HE multiuser (MU) PPDU for DL MU-MIMO or MU-OFDMA;
4. HE trigger based (TB) PPDU for UL MU-MIMO or MU-OFDMA.

All four PPDU formats start with a legacy PLCP preamble and header for backward compatibility. The legacy portion is followed by RL-SIG field, which is a repeat of L-SIG and an 8-$\mu$s long HE-SIG-A field. The HE-SIG-A field specifies the PHY parameters required to decode the rest of the PPDU. In a HE ER SU PPDU, the HE-SIG-A duration is doubled by duplicating the contents for redundancy to aid long-range use cases. The HE-SIG-A is followed by a variable length HE-SIG-B field that is present only in HE MU PPDU and it contains per user information such as RU and stream allocation for all users that are part of the MU transmission. Up until HE-SIG-B, only pre-HE modulation is employed. This is followed by HE-STF and HE-LTF fields that are used for MIMO AGC and MIMO channel estimation. The HE-STF is 8-$\mu$s long in HE TB PPDU, while it is 4-$\mu$s long for other PPDU types. The number of HE-LTFs depends on the total number of spatial streams in the PPDU. For example, a MU PPDU carrying two spatial streams each for two users will have four HE-LTFs. The HE-LTFs are followed by the payload or PSDU that is transmitted at HE data rates. The PPDU ends with a packet extension (PE) field, which provides additional time for low complexity receiver implementations to decode the PPDU and prepare a response frame. The minimum PE requirement of a device is advertised in the PPE thresholds field of HE capabilities information element (IE) as shown in Figure 3.2.

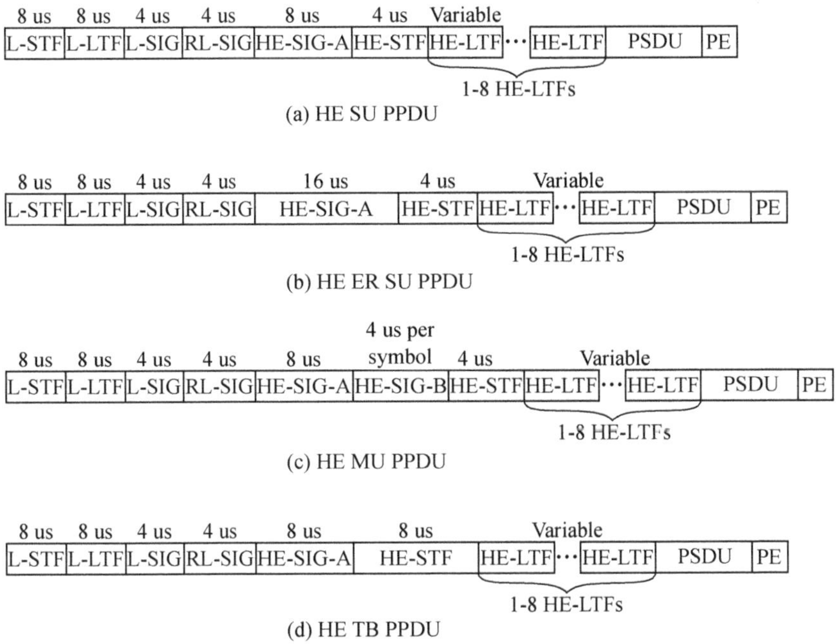


**Figure 3.1** HE PPDU formats.

| 8b | 8b | 8b | 48b | 88b | 32b, 64b or 96b | variable |
|---|---|---|---|---|---|---|
| Element ID | Length | Element ID Extension | HE MAC Capabilities Information | HE PHY Capabilities Information | Supported HE-MCS And NSS Set | PPE Thresholds |

**Figure 3.2** HE capabilities IE.

### 3.1.1 Long OFDM Symbol

Wi-Fi 6 quadruples both the OFDM symbol duration and the number of subcarriers to improve the PHY layer efficiency and also provide increased flexibility for OFDMA RU allocation. Each HE OFDM symbol is 12.8 $\mu$s long and has a subcarrier spacing of 78.125 kHz.

### 3.1.2 GI and HE LTF

Wi-Fi 6 defines three different GIs (0.8 $\mu$s, 1.6 $\mu$s, 3.2 $\mu$s) and three different HE-LTF durations (1x : 3.2 $\mu$s, 2x : 6.4 $\mu$s, 4x : 12.8 $\mu$s), but only four GI and HE-LTF duration combinations, as specified in Table 3.1, are allowed. For SU, ER SU, and MU PPDUs, the HE-SIG-A field specifies the GI and LTF size employed in that PPDU while it is specified by a preceding trigger frame for HE TB PPDUs.

While 0.8-$\mu$s GI is the most commonly used GI, the higher GIs are optimal for outdoor applications, where the wireless channels tend to have larger delay spreads. A longer HE-LTF duration provides better channel estimation accuracy at the receiver. The multiple options for HE-LTF duration enable the transmitter to tradeoff between channel estimation accuracy and the associated HE-LTF overhead on a per PPDU basis.

**Table 3.1**

GI and HE-LTF Sizes

| *GI+LTF* | *HE SU or HE ER SU* | | *HE MU* | | *HE TB* | |
|---|---|---|---|---|---|---|
| | *GI* | *LTF* | *GI* | *LTF* | *GI* | *LTF* |
| 0 | 0.8 $\mu$s | 1x | 0.8 $\mu$s | 4x | 1.6 $\mu$s | 1x |
| 1 | 0.8 $\mu$s | 2x | 0.8 $\mu$s | 2x | 1.6 $\mu$s | 2x |
| 2 | 1.6 $\mu$s | 2x | 1.6 $\mu$s | 2x | 3.2 $\mu$s | 4x |
| 3 | 3.2 $\mu$s | 4x | 3.2 $\mu$s | 4x | reserved | |

## 3.2 SINGLE USER OPERATION

Wi-Fi 6 has several subtle but important enhancements to SU operation that improve efficiency and throughput across all signal-to-noise ratio (SNR).

### 3.2.1 Coding and Modulation

Wi-Fi 6 supports modulations ranging from BPSK to 1024-QAM, with 1024-QAM being a new addition. The supported NSS and bandwidth (BW) modes in Wi-Fi 6 are identical to Wi-Fi 5. While Wi-Fi 6 supports both low density parity check (LDPC) code and binary convolutional code (BCC), only LDPC coding is allowed for high data rates involving NSS $>$ 4 or 1024-QAM modulation or BW greater than 20 MHz. Wi-Fi 6 has lower GI overhead as measured by (GI duration / OFDM symbol duration) and higher data subcarrier allocation as measured by (Number of data subcarriers / Total number of subcarriers) when compared to Wi-Fi 5 as shown in Table 3.2. For the same modulation and coding scheme (MCS), spatial stream, and BW, the SU data rate for Wi-Fi 6 with 0.8-$\mu$s GI is at least 10% higher than Wi-Fi 5 with 0.4-$\mu$s GI. The HE SU data rates for 1 spatial stream and 0.8-$\mu$s GI are listed in Table 3.3 for reference.

**Table 3.2**
HE Vs VHT SU Subcarrier Allocation

| | *20 MHz* | | *40 MHz* | | *80 MHz* | |
|---|---|---|---|---|---|---|
| | *VHT* | *HE* | *VHT* | *HE* | *VHT* | *HE* |
| Total subcarriers | 64 | 256 | 128 | 512 | 256 | 1024 |
| Data subcarriers | 52 | 234 | 108 | 468 | 234 | 980 |
| Pilot subcarriers | 4 | 8 | 6 | 16 | 8 | 16 |
| Unused subcarriers | 8 | 14 | 14 | 28 | 14 | 28 |
| Data subcarriers / Total subcarriers | 81.25% | 91.4% | 84.3% | 91.4% | 91.4% | 95.7% |

### 3.2.2 Dual Carrier Modulation

Dual carrier modulation (DCM) is yet another new feature that can help increase robustness of the wireless link especially at far range. This is achieved by modulating two subcarriers with the same information and this redundancy can be exploited by the receiver to reduce decoding errors. DCM can potentially mitigate any narrowband interference. DCM reduces the data rate by half and is supported only for MCS 0, 1, 3, 4, and NSS less than 3. DCM can be applied on either HE SU PDDU or HE ER SU PPDU and the presence of DCM is indicated in the HE-SIG-A field.

### 3.2.3 HE TxBf and Channel Sounding

Transmit beamforming (TxBf) in Wi-Fi 6 is mostly similar to Wi-Fi 5. One of the differences though is that the HE compressed beamforming (CBF) report supports subcarrier grouping of only $N_g = 4$ or 16. Since Wi-Fi 6 supports four times higher number of subcarriers, the increased subcarrier grouping helps keep the frequency resolution of channel state information same as in Wi-Fi 5. The HE CBF report size is therefore only incrementally higher than VHT owing to the higher

**Table 3.3**
HE SU Data Rates in Mbps for 1 Spatial Stream

| *Modulation* | *Coding Rate* | *MCS* | *GI (μs)* | *20 MHz* | *40 MHz* | *80 MHz* | *160 MHz* |
|---|---|---|---|---|---|---|---|
| BPSK | 1/2 | 0 | 0.8 | 8.6 | 17.2 | 36 | 72.1 |
| BPSK | 1/2 | 0 | 1.6 | 8.1 | 16.3 | 34 | 68.1 |
| BPSK | 1/2 | 0 | 3.2 | 7.3 | 14.6 | 30.6 | 61.3 |
| QPSK | 1/2 | 1 | 0.8 | 17.2 | 34.4 | 72.1 | 144.1 |
| QPSK | 1/2 | 1 | 1.6 | 16.3 | 32.5 | 68.1 | 136.1 |
| QPSK | 1/2 | 1 | 3.2 | 14.6 | 29.3 | 61.3 | 122.5 |
| QPSK | 3/4 | 2 | 0.8 | 25.8 | 51.6 | 108.1 | 216.2 |
| QPSK | 3/4 | 2 | 1.6 | 24.4 | 48.8 | 102.1 | 204.2 |
| QPSK | 3/4 | 2 | 3.2 | 21.9 | 43.9 | 91.9 | 183.8 |
| 16-QAM | 1/2 | 3 | 0.8 | 34.4 | 68.8 | 144.1 | 288.2 |
| 16-QAM | 1/2 | 3 | 1.6 | 32.5 | 65 | 136.1 | 272.2 |
| 16-QAM | 1/2 | 3 | 3.2 | 29.3 | 58.5 | 122.5 | 245 |
| 16-QAM | 3/4 | 4 | 0.8 | 51.6 | 103.2 | 216.2 | 432.4 |
| 16-QAM | 3/4 | 4 | 1.6 | 48.8 | 97.5 | 204.2 | 408.3 |
| 16-QAM | 3/4 | 4 | 3.2 | 43.9 | 87.8 | 183.8 | 367.5 |
| 64-QAM | 2/3 | 5 | 0.8 | 68.8 | 137.6 | 288.2 | 576.5 |
| 64-QAM | 2/3 | 5 | 1.6 | 65 | 130 | 272.2 | 544.4 |
| 64-QAM | 2/3 | 5 | 3.2 | 58.5 | 117 | 245 | 490 |
| 64-QAM | 3/4 | 6 | 0.8 | 77.4 | 154.9 | 324.3 | 648.5 |
| 64-QAM | 3/4 | 6 | 1.6 | 73.1 | 146.3 | 306.3 | 612.5 |
| 64-QAM | 3/4 | 6 | 3.2 | 65.8 | 131.6 | 275.6 | 551.3 |
| 64-QAM | 5/6 | 7 | 0.8 | 86 | 172.1 | 360.3 | 720.6 |
| 64-QAM | 5/6 | 7 | 1.6 | 81.3 | 162.5 | 340.3 | 680.6 |
| 64-QAM | 5/6 | 7 | 3.2 | 73.1 | 146.3 | 306.3 | 612.5 |
| 256-QAM | 3/4 | 8 | 0.8 | 103.2 | 206.5 | 432.4 | 864.7 |
| 256-QAM | 3/4 | 8 | 0.8 | 97.5 | 195 | 408.3 | 816.7 |
| 256-QAM | 3/4 | 8 | 0.8 | 87.8 | 175.5 | 367.5 | 735 |
| 256-QAM | 5/6 | 9 | 0.8 | 114.7 | 229.4 | 480.4 | 960.7 |
| 256-QAM | 5/6 | 9 | 1.6 | 108.3 | 216.7 | 453.7 | 907.4 |
| 256-QAM | 5/6 | 9 | 3.2 | 97.5 | 195 | 408.3 | 816.6 |
| 1024-QAM | 3/4 | 10 | 0.8 | 129 | 258.1 | 540.4 | 1080.9 |
| 1024-QAM | 3/4 | 10 | 1.6 | 121.9 | 243.8 | 510.4 | 1020.8 |
| 1024-QAM | 3/4 | 10 | 3.2 | 109.7 | 219.4 | 459.4 | 918.8 |
| 1024-QAM | 5/6 | 11 | 0.8 | 143.4 | 286.8 | 600.4 | 1201 |
| 1024-QAM | 5/6 | 11 | 1.6 | 135.4 | 270.8 | 567.1 | 1134.2 |
| 1024-QAM | 5/6 | 11 | 3.2 | 121.9 | 243.8 | 510.4 | 1020.8 |

fraction of data subcarriers. The number of bits in HE CBF report is given by $8 \cdot N_c + (N_s \cdot N_a \cdot \text{bit precision})$ bits, where $N_s$ is the number of subcarriers for which channel feedback is reported and can be obtained from Table 3.4. The number of angles $N_a$ is derived from $N_r$, $N_c$ as in Wi-Fi 5. The supported average bit precision for the angles are 3 or 5 bits for SU feedback and 6 or 8 bits for MU feedback.

**Table 3.4**

$N_s$ in HE CBF Report for Different BW, $N_g$

| *BW (MHz)* | $N_g$ | $N_s$ |
|---|---|---|
| 20 | 4 | 64 |
| 20 | 16 | 20 |
| 40 | 4 | 122 |
| 40 | 16 | 32 |
| 80 | 4 | 250 |
| 80 | 16 | 64 |
| 160 | 4 | 502 |
| 160 | 16 | 128 |

Wi-Fi 6 supports a new option for beamformer to request via NDPA frame a channel quality indication (CQI) only report from beamformee instead of the complete channel state information. The CQI only report is a small report that contains a number representing channel quality (for example SNR or condition number) for each spatial stream, which can assist the SU beamformer to decide $N_c$ for subsequent channel sounding or the optimum NSS for data traffic, thus reducing time spent on trying different number of spatial streams (NSS). Being a new feature, the CQI only report is not yet commonly used in practice. A new HE NDPA frame as shown in Figure 3.3 replaces the VHT NDPA frame to enable these enhancements.

### 3.2.4 Midamble

Midamble is a new Wi-Fi 6 feature that enables use cases involving moderate mobility at vehicular speeds of up to 60 kilometers per hour. An example of such an use case could be wireless control of drones using Wi-Fi. Midambles are additional HE-LTF symbols that are inserted every 10 or 20 OFDM symbols in the payload portion of a physical layer conformance procedure protocol data unit (PPDU) to facilitate a receiver to update its channel estimate. The presence of midamble and

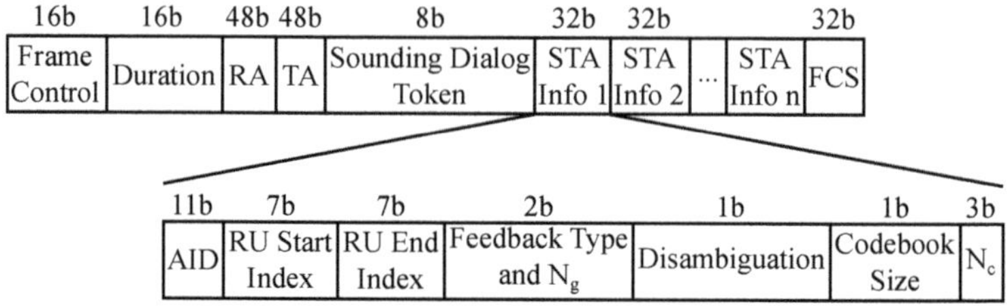


**Figure 3.3** HE NDPA frame format.

the periodicity of midamble are indicated in the HE-SIG-A field. Although this feature has a good potential in certain use cases, Wi-Fi 6 application specific integrated circuits (ASICs) in the market today are yet to support this optional feature.

### 3.2.5 Multi-TID AMPDU

Multi-TID AMPDU is a new type of AMPDU aggregation that allows aggregation of MPDUs corresponding to multiple TIDs. The recipient of a multi-TID AMPDU can acknowledge it using a newly introduced multi-STA blockack (BA) frame, which is described in detail in Section 3.3.3.2. A multi-STA BA frame in general can be used to acknowledge MPDUs from different STAs using per AID TID info fields, but in the context of SU traffic, a multi-STA BA frame can aggregate BAs to the same recipient for multiple TIDs. The acknowledgment for a multi-TID AMPDU can also be solicited using a new multi-TID BAR frame. In order to maintain quality of service (QoS) prioritization, a multi-TID AMPDU can contain MPDUs only from TIDs having the same or higher priority than the primary access category, where primary access category refers to the access category used to gain medium access as per EDCA.

### 3.2.6 Blockack Length

Wi-Fi 6 provides multiple options for BA bitmap lengths (32, 64, 128, 256). The maximum BA length used in a BA session depends on the negotiated buffer size parameter during ADDBA request/response exchange as described in Table 3.5. A device can choose an appropriate BA length within a BA session depending on the difference in sequence number between last received and last acknowledged MPDU, or other metrics based on traffic profile.

**Table 3.5**
Blockack Bitmap Lengths

| | Buffer Size | | |
|---|---|---|---|
| | *1 to 64* | *65 to 128* | *129 to 256* |
| BA bitmap length for BA frame | 64 | 64 or 256 | 64 or 256 |
| BA bitmap length for multi-STA BA frame | 32 or 64 | 32, 64 or 128 | 32, 64, 128 or 256 |

### 3.2.7 Aggregation Limits

The maximum MSDU length, MPDU length, and PPDU duration for HE are the same as for VHT. However, owing to the higher 256-bit BA bitmap length, HE allows a four times higher aggregated MAC protocol data unit (AMPDU) aggregation limit compared to VHT. The maximum AMPDU length for HE PPDUs is determined by the maximum A-MPDU length exponent field in VHT capabilities IE along with the maximum A-MPDU length exponent extension subfield of HE MAC capabilities information field in HE capabilities IE and can be up to 4M bytes.

### 3.2.8 Duration-based RTS-CTS

One of the issues with RTS-CTS protection in earlier Wi-Fi generations is that the AP cannot influence or control the STA's usage of RTS-CTS protection. Since RTS usage is not mandatory and each client type has its own logic to decide RTS protection, a network cannot reap the full benefits of RTS-CTS. Moreover, a STA does not know certain information such as the number of active and associated STAs in the network to predict the collision probability. An AP is in a better position to advise to STAs when it is appropriate to use RTS-CTS protection. Wi-Fi 6 addresses this problem by introducing a mechanism for an AP to control the RTS usage of associated STAs. A Wi-Fi 6 AP can use the TxOP duration RTS threshold subfield of HE operation parameters field in the HE operation IE to indicate to associated Wi-Fi 6 STAs when to apply RTS protection. Figure 3.4 shows the format of HE operation IE. The TxOP duration RTS threshold subfield can take on any value between 1 and 1023. A value of 1023 implies RTS usage is not required. Any value less than 1023 implies all associated Wi-Fi 6 STAs are required to use RTS protection for any unicast data transmission that occupies the medium for TxOP duration $> 32 \cdot$ (TxOP duration RTS threshold) $\mu$s.

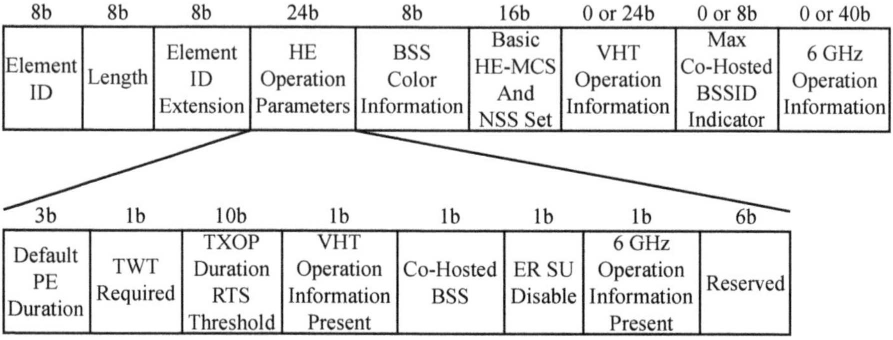


**Figure 3.4** HE operation IE.

## 3.3 MULTIUSER OPERATION

While Wi-Fi 6 has several improvements to SU efficiency, the major network efficiency improvement of Wi-Fi 6 comes from MU operation. Wi-Fi 6 supports MU operation for both DL and UL traffic using MU-MIMO or MU-OFDMA or a combination of both. While MU-MIMO is possible only if the AP has more antennas than the STAs, MU-OFDMA is possible on all Wi-Fi 6 APs regardless of the number of antennas.

### 3.3.1 OFDMA Resource Unit Allocation

A resource unit (RU) is a group of consecutive subcarriers or tones that can be allocated to different users in MU-OFDMA scheduling. While a majority of the tones in each RU is used for data, a small fraction of tones is meant for pilots. Also, a small fraction of tones is not part of any RU to serve as null subcarriers or guard bands. Wi-Fi 6 defines six different RUs that vary in size: 996-tone RU, 484-tone RU, 242-tone RU, 106-tone RU, 52-tone RU, and 26-tone RU. The 26-tone RU is the smallest RU size and it occupies approximately 2-MHz BW. The 242-tone RU, 484-tone RU, and 996-tone RU in 20-MHz, 40-MHz, and 80-MHz BW modes occupy the full channel BW for one user, and they match the same data and pilot subcarrier allocation as for SU. Table 3.6 shows the data and pilot tone allocation for different RUs.

**Table 3.6**
Data and Pilot Subcarriers for Different RUs

| *RU Size* | *Data Subcarriers* | *Pilot Subcarriers* |
|---|---|---|
| 26-tone | 24 | 2 |
| 52-tone | 48 | 4 |
| 106-tone | 102 | 4 |
| 242-tone | 234 | 8 |
| 484-tone | 468 | 16 |
| 996-tone | 980 | 16 |

The RU allocations possible in Wi-Fi 6 have some stringent constraints due to the fixed position of the RUs as shown in Figure 3.5 for 20-MHz, 40-MHz, and 80-MHz BW. The 160-MHz RU locations follow the same structure as 80 MHz in both the primary 80 MHz and secondary 80-MHz subchannels. The valid RU allocations in Wi-Fi 6 can be easily understood by the following rules:

- A 996-tone RU can be replaced by two 484-tone RUs, two unused tones, and one 26-tone RU in the middle.
- A 484-tone RU can be replaced by two 242-tone RUs.
- A 242-tone RU can be replaced by two 106-tone RUs, four unused tones, and one 26-tone RU in the middle.
- A 106-tone RU can be replaced by two 52-tone RUs and two unused tones.
- A 52-tone RU can be replaced by two 26-tone RUs.

Table 3.7 shows the maximum number of users and data subcarrier efficiency for some RU allocations. Although the standard permits to use the center 26-tone RU in 20-MHz and 80-MHz BW modes, most implementations do not allocate it due to interference from DC leakage. The absence of the center 26-tone RU reduces the data subcarrier efficiency to 74% and 84%, respectively, in 20-MHz and 80-MHz BW.

### 3.3.2 Downlink MU Operation

The frame exchanges involved in DL MU operation are shown in Figure 3.6. DL MU transmission employs a HE MU PPDU, which aggregates PSDUs destined

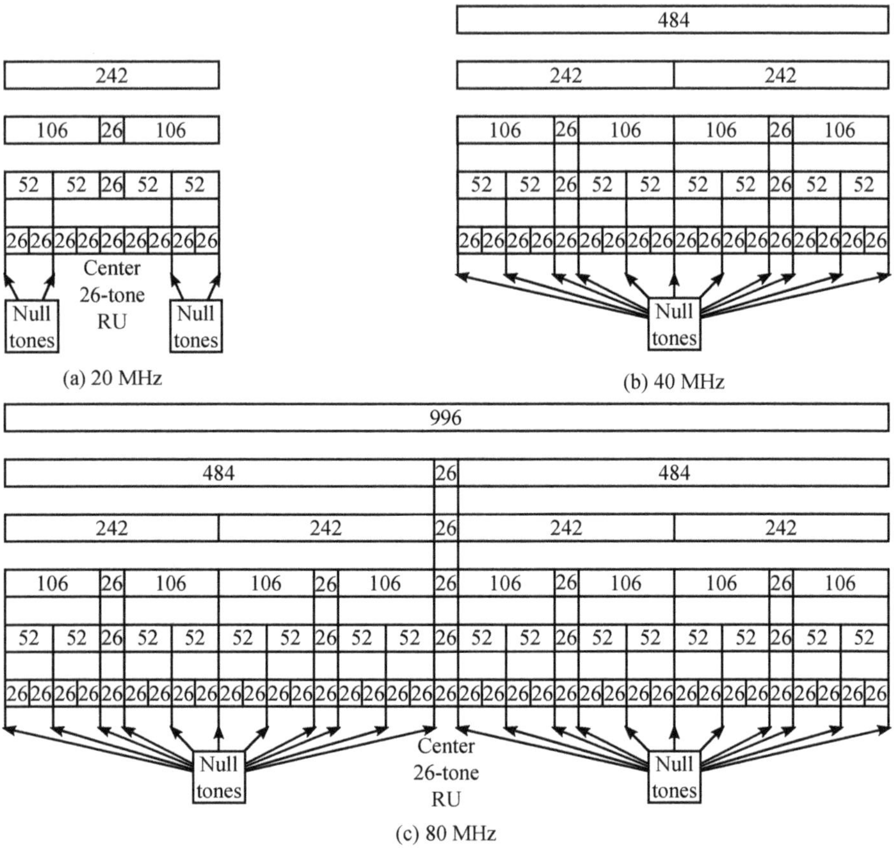


**Figure 3.5** RU locations in 20-, 40- and 80-MHz BW.

for different STAs. The PSDUs for different STAs are transmitted simultaneously either on different spatial streams or on different RUs. The AIDs of the intended recipient STAs and their respective stream or RU allocation per STA is present in the HE-SIG-B field of HE MU PPDU. This enables the AP to decide on a per MU PPDU basis, the group of STA recipients and their respective stream or RU allocation. Although the standard allows combining MU-MIMO and MU-OFDMA in the same MU PPDU, most implementations keep it simple by choosing either MU-MIMO or MU-OFDMA. One advantage of MU-OFDMA over MU-MIMO is

**Table 3.7**
Maximum Number of Users and Data Subcarrier Efficiency for Different RU Allocation

| *BW* (MHz) | *RU Allocation* | *Number of Users* | *Number of Data Subcarriers* | *Data Subcarrier Efficiency* |
|---|---|---|---|---|
| 20 | 1x 242-tone RU | 1 | 234 | 91.4% |
| 20 | 2x 106-tone RU, 1x center 26-tone RU | 3 | 228 | 89.0% |
| 20 | 4x 52-tone RU, 1x center 26-tone RU | 5 | 216 | 84.3% |
| 20 | 8x 26-tone RU, 1x center 26-tone RU | 9 | 216 | 84.3% |
| 40 | 1x 484-tone RU | 1 | 468 | 91.4% |
| 40 | 2x 242-tone RU | 2 | 468 | 91.4% |
| 40 | 4x 106-tone RU, 2x 26-tone RU | 6 | 456 | 89.0% |
| 40 | 8x 52-tone RU, 2x 26-tone RU | 10 | 432 | 84.3% |
| 40 | 18x 26-tone RU | 18 | 432 | 84.3% |
| 80 | 996-tone RU | 1 | 980 | 95.7% |
| 80 | 2x 484-tone RU, 1x center 26-tone RU | 2 | 962 | 93.9% |
| 80 | 4x 242-tone RU, 1x center 26-tone RU | 5 | 962 | 93.9% |
| 80 | 8x 106-tone RU, 4x 26-tone RU, 1x center 26-tone RU | 13 | 936 | 91.4% |
| 80 | 16x 52-tone RU, 4x 26-tone RU, 1x center 26-tone RU | 21 | 888 | 86.7% |
| 80 | 36x 26-tone RU, 1x center 26-tone RU | 37 | 888 | 86.7% |

it doesn't require any prior channel sounding. However, if TxBf is desired on top of MU-OFDMA, then channel sounding is needed. The number of STAs in an MU PPDU is constrained by the maximum number of RUs in the case of MU-OFDMA and by the maximum stream capability of the AP in the case of MU-MIMO.

### 3.3.3 Uplink MU Operation

When a large number of clients are connected to an AP, a significant fraction of airtime is wasted in collisions leading to inefficiency. Figure 3.7 shows simulation results of the collision probability in a network as a function of the number of devices contending for the medium. These simulation results assume best effort access category traffic and absence of hidden nodes in the network. Although the use of RTS-CTS protection can help with hidden nodes and reduce the airtime loss due to collisions, it cannot eliminate collisions. A more efficient solution is uplink MU, which enables multiple STAs to transmit simultaneously on either different RUs

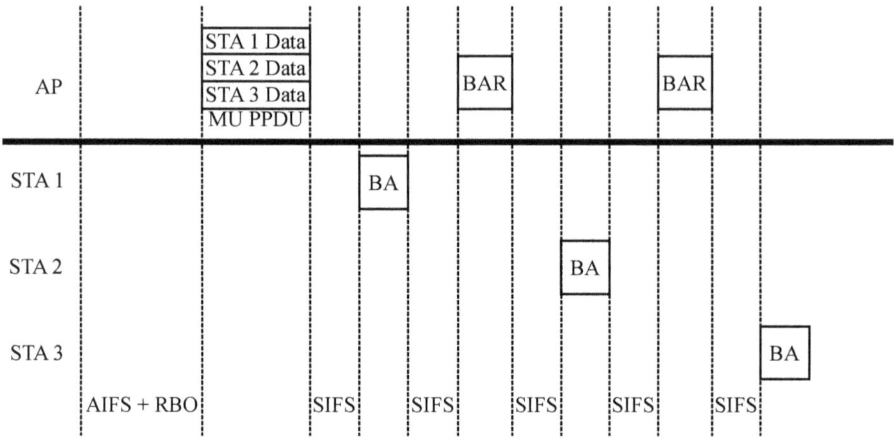


**Figure 3.6** Downlink MU operation.

or different spatial streams. Such scheduled uplink MU transmissions can reduce collisions amongst STAs thereby minimizing the medium access latency.

UL MU operation using either the MU-MIMO or MU-OFDMA technique is activated in Wi-Fi 6 by a newly introduced trigger frame transmitted by the AP. This frame acts as a timing trigger for soliciting the intended STA participants to transmit their data at the same time using TB PPDU. The transmissions from multiple STAs have to be synchronized in symbol timing and carrier frequency for the AP to decode all the STA's data using simple signal processing techniques. In UL MU-OFDMA, each STA's data is present in different RUs that do not interfere with each other. UL MU-MIMO, on the other hand, places each STA's data on different spatial streams and the AP jointly decodes the spatial streams of all users using the same methods as in SU MIMO reception. In SU MIMO, all the spatial streams originate from one device, so the OFDM symbols for all spatial streams are naturally aligned in time. But, in UL MU-MIMO, this timing alignment of OFDM symbols from the different STAs is achieved using the trigger frame. One major advantage of UL MU-MIMO over DL MU-MIMO is that there is no MU sounding overhead involved because the HE-LTFs in the TB PPDU enable the AP to estimate the channel between each involved STA and AP. Furthermore, the STAs that are part of UL MU-MIMO do not have to apply any special steering mechanism like null-steering as in the case of DL MU-MIMO. Essentially, UL MU-MIMO offers the spatial multiplexing gains of DL MU-MIMO without the additional sounding

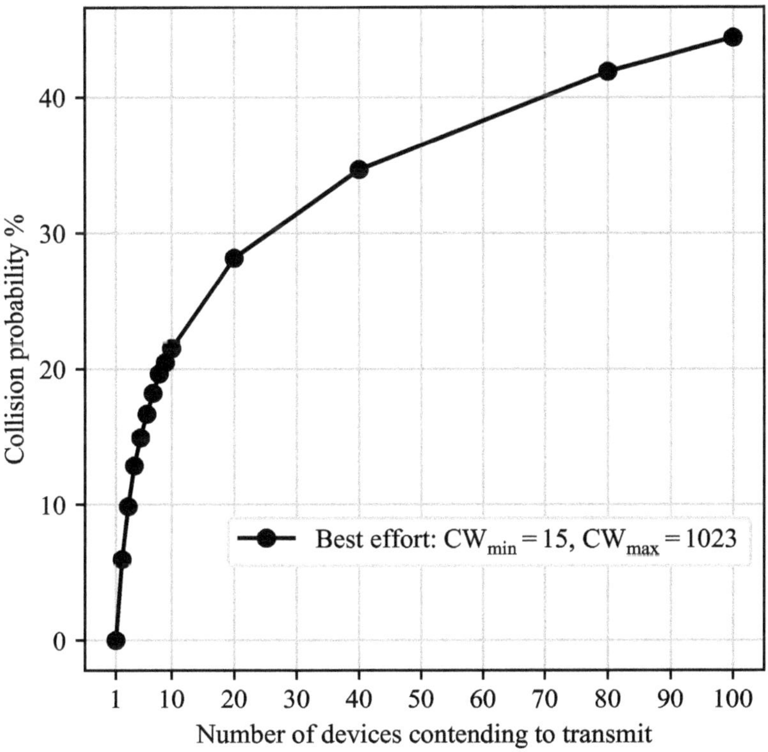


**Figure 3.7** Collision probability.

overhead. Figure 3.8 shows the timing diagram of the different frames involved in UL MU operation.

3.3.3.1 Trigger Frame

The trigger frame is a control frame (subtype value of 0010 in frame control field of medium access control (MAC) header), which contains AIDs of the intended STA participants and their respective spatial stream or RU allocation. Upon receiving the

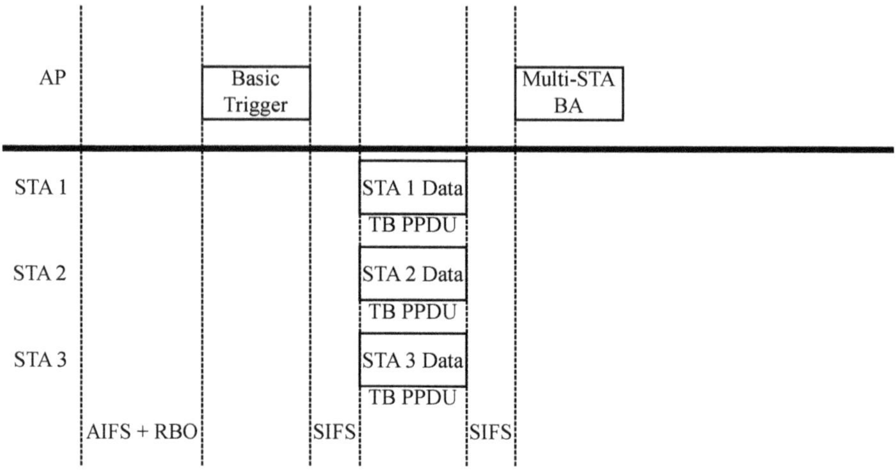


**Figure 3.8** Uplink MU operation.

trigger frame, the STAs whose AID match the ones specified in the trigger frame are required to transmit their data in their allocated spatial stream or RU using a HE TB PPDU in exactly SIFS time interval. The standard requires the timing accuracy error and carrier frequency offset (CFO) error on the TB PPDU to be below 0.4 $\mu$s and 350 Hz, respectively. The minimum OFDM GI permitted in a TB PPDU is 1.6 $\mu$s to allow for different multipath delays between the STAs.

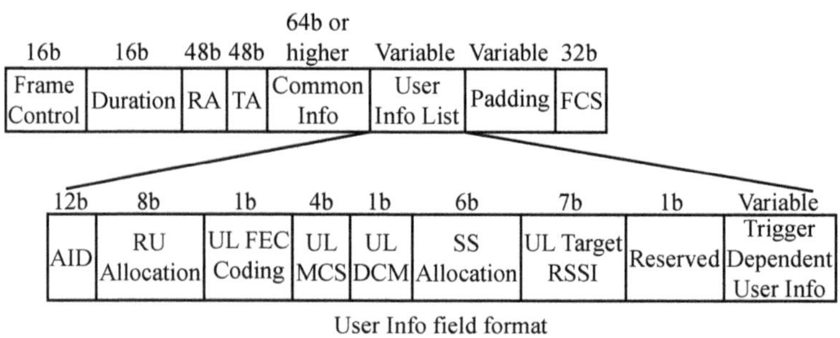


**Figure 3.9** Trigger frame format.

Figure 3.9 details the various fields in a trigger frame. The AP can transmit a trigger frame in legacy, HT, VHT, or HE data rate. The common info field of trigger frame specifies the trigger frame type, Tx power of the trigger frame in dBm and several key parameters of the following TB PPDU such as the L-SIG length field, BW, GI, and HE LTF. The trigger frame type used by AP for receiving QoS data with UL MU operation is called basic trigger. There are several other trigger types for varied purposes described in detail in Section 3.3.4. The user info list field contains a user info field per participating STA. The user info field specifies the AID, RU allocation, spatial stream allocation, and MCS to be used by each participating STA in the TB PPDU. It also contains a target RSSI subfield which controls the Tx power for the TB PDDU. Upon receiving the trigger frame, each participating STA first computes the path loss between the AP and STA by subtracting the Rx RSSI in dBm from the Tx power mentioned in the Common Info field. Then, each participating STA determines the Tx power for the TB PPDU by adding the computed path loss to the target RSSI specified in the user info field. This UL transmit power control mechanism allows the AP to ensure that the TB PPDUs from all participating STA arrive at a RSSI level within the AP receiver's dynamic range. The trigger dependent user info subfield in a basic trigger frame can be used by the AP to specify a preferred access category (AC) per STA. The preferred AC sets the minimum priority AC traffic that can be sent by participating STA. In summary, the AP determines the list of participating STAs, along with the BW, MCS, RU allocation, spatial stream allocation, Tx power, preferred AC, and maximum duration of the TB PPDU per participating STA.

Trigger frames can also be piggybacked in a MU PPDU as part of an AMPDU to solicit uplink (UL) MU immediately following downlink (DL) MU. Alternatively, UL MU can be solicited using the triggered response scheduling (TRS) control subfield of HT control field in an MPDU of a MU PPDU. The TRS control subfield essentially plays the role of trigger frame and can be used only with STAs that support this optional capability. Support for TRS control subfield is indicated by the TRS support field of HE MAC capabilities IE.

#### 3.3.3.2 Multi-STA Blockack

Upon reception of a TB PPDU, the AP acknowledges all successfully received MPDUs in it using a single multi-STA BA frame in SIFS time. This avoids the overhead of multiple SIFS + BAR-BA exchanges with each individual STA as in the case of DL MU operation. Figure 3.10 shows the frame format of a multi-STA BA frame. The BA information field in multi-STA BA essentially aggregates the BA information; namely, starting sequence number (SSN) and BA bitmap for all participating STAs. The RA field is set to broadcast MAC address and the participating STAs extract their respective BA information using the AID subfield in per AID TID info subfield of BA information field.

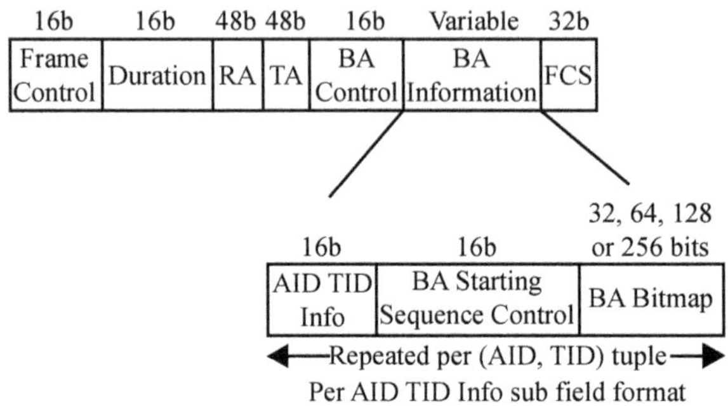


**Figure 3.10** Multi-STA BA frame format.

### 3.3.4 Trigger Frame Types

There are multiple trigger frame types defined in Wi-Fi 6 to enable UL MU for non-QoS data frames such as BA and beamforming report. Table 3.8 details the different trigger types along with a brief description of their purpose.

#### 3.3.4.1 MU BAR

MU BAR trigger type can be used by an AP to efficiently solicit BAs in a DL MU operation, which is one of the significant overhead causes of DL MU. The trigger dependent user info subfield in a MU BAR trigger frame specifies the SSN per STA, which in this case corresponds to the last acknowledged sequence number by the STA. Each STA uses this information to prepare the BA for the appropriate window of sequence numbers in a TB PPDU. As in basic trigger, the user info list field specifies the AIDs of STAs and their respective RU or spatial stream allocation for the TB PPDU. DL MU combined with UL MU for the BAs considerably improves the efficiency especially for DL MU-OFDMA, since it may involve a larger number of STAs. In practice, several implementations piggyback the MU BAR trigger frame as part of an AMPDU in a MU PPDU.

**Table 3.8**
Trigger Frame Types

| *Trigger Type Subfield Value* | *Trigger Frame Type* | *Purpose* |
|---|---|---|
| 0 | Basic | UL MU for QoS data. |
| 1 | Beamforming report poll (BFRP) | UL MU for retrieving beamforming report from multiple STAs. |
| 2 | MU BAR | UL MU for retrieving BA from multiple STAs. |
| 3 | MU RTS | RTS-CTS protection for MU operation. |
| 4 | Buffer status report poll | UL MU for AP to query the amount of queued traffic at multiple STAs. |
| 5 | GCR MU BAR | MU BAR specifically for retrieving BA from STAs of a multicast group. |
| 6 | BW query report poll | UL MU for AP to query STA's CCA status per 20-MHz subchannel. |
| 7 | NDP feedback report poll | For AP to query resource request and approximate queue size from STAs using UL MU. |

#### 3.3.4.2 GCR MU BAR

One of the problems with multicast over Wi-Fi is there is no acknowledgment available from the STAs of a multicast group. Wi-Fi 6 provides a new option called group cast with retries (GCR) for performing multicast transmission to Wi-Fi 6 clients potentially improving the multicast performance. In this method, AP forms a multicast group and sends a multicast frame to the STA group members. This is followed by a GCR MU BAR trigger frame, which solicits BAs from group members. Based on the BA contents from each group member, AP can perform multicast rate adaptation and retransmit only the failed MPDUs, very similar to SU operation. The ability to obtain acknowledgments from group members opens up the possibility to transmit multicast data at high PHY rates, thereby improving multicast performance.

#### 3.3.4.3 Beamforming report poll

A beamforming report poll (BFRP) trigger frame can be applied in a MU sounding sequence to solicit beamforming report from multiple STAs using UL MU. This method of MU sounding is called HE TB sounding sequence, which is depicted in Figure 3.11. A HE TB sounding sequence begins with a HE NDPA frame that specifies the list of STAs that are part of the TB sounding sequence and also specifies the various parameters for the channel feedback such as feedback type, subcarrier grouping, and bit precision. In SIFS time, it is followed by a NDP frame that is used by the STAs to measure the channel and prepare the CBF report. The NDP is followed in SIFS time by the BFRP trigger frame that solicits CBF report from the STAs. The STAs then respond to the BFRP trigger frame in SIFS time with a TB PPDU containing their respective CBF reports. The HE TB sounding sequence and MU blockack request (BAR) are two enhancements that improve the efficiency of DL multiuser MIMO (MU-MIMO) feature in Wi-Fi 6 as compared to Wi-Fi 5.

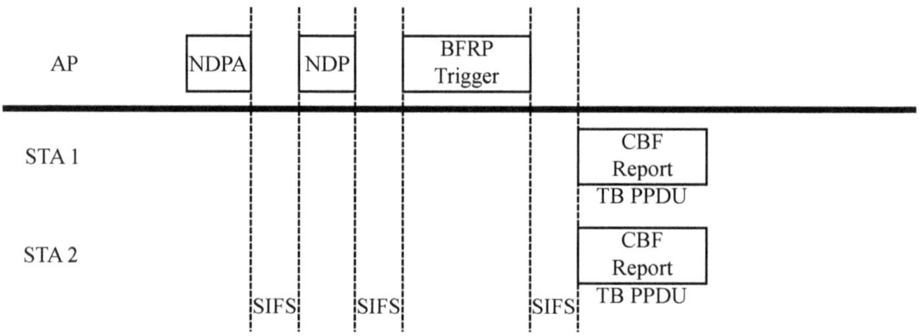


**Figure 3.11** HE TB sounding sequence.

#### 3.3.4.4 MU RTS

The MU RTS trigger type offers a mechanism to provide collision protection from hidden nodes for DL or UL MU sequence with Wi-Fi 6 clients. Unlike other trigger types, the MU RTS frame does not solicit a TB PPDU in response and therefore does not contain any RU or spatial stream allocation. The MU RTS is always sent by the AP in legacy PHY PPDU format and is duplicated across all subchannels for 40 MHz or higher BW, so that all legacy devices can decode the duration field in the MAC header. The duration field in MU RTS sets the network allocation vector (NAV) in all neighboring devices close to the AP and the AP assigns the duration field value such that the entire MU sequence is protected. The MU RTS has the TA field set to the AP's MAC address and the RA field set to the broadcast address. The user info list field of MU RTS contains the list of STA AIDs that are to participate in the MU sequence. However, this information will be understood only by participating Wi-Fi 6 STAs. In response to the MU-RTS, all the STAs with matching AID are required to respond with a regular CTS frame at 6-Mbps legacy PHY PPDU format in SIFS time. The RA field in the CTS frame is set to the AP's MAC address and the TA field is not present. Since the CTS frame has no TA field, the contents of the CTS frame sent by all the participating Wi-Fi 6 STAs are exactly identical. From the point of view of any Wi-Fi device in the vicinity, it receives multiple copies of the same CTS frame from different STAs with possibly different multipath delays. As long as the difference between the smallest and largest multipath delay is within the 0.8-$\mu$s GI, any legacy Wi-Fi device can decode the CTS frame contents. The duration field in the CTS frame thus sets the NAV in all neighboring devices close to all the participating STAs. Hence, the DL or UL MU sequence that follows the MU RTS-CTS exchange is protected from collisions. One subtle point to note here is that the AP does not know which STAs responded with CTS as there is no TA field in CTS frame. As long as AP receives a CTS from even one of the participating STAs it will proceed with the MU sequence, which allows for a small chance of collision. However, MU RTS still offers reasonable collision protection with less overhead. Figure 3.12 illustrates a DL MU sequence with MU RTS protection and MU BAR trigger piggybacked on MU PPDU to retrieve BAs using UL MU.

### 3.3.5 Buffer Status Report

One of the challenges with UL MU is the UL MU scheduler at the AP, which schedules STAs for UL MU and allocates resource. In order to perform an effective and fair UL MU resource allocation, the AP needs to know the amount of buffered traffic or queue size at the STA per TID. There are three methods for an AP to obtain a buffer status report (BSR) from the STA. In the first method, STA sends unsolicited BSR to the AP in the QoS control field of any QoS null or QoS data frame. In this method, a STA would require 8 QoS null or QoS data frames to send the BSR information for 8 TIDs, but this method provides precise queue size (QS) information per TID. The AP can identify the presence of BSR if bit 4 of QoS control field is set to 1. The buffer status is represented by scaling factor subfield (bits 14-15 of QoS control) and unscaled value subfield (bits 8-13 of QoS control). The scaling factor subfield can be mapped to scaling factor ($SF$) value in bytes using (3.1). Table 3.9 shows how a STA encodes the UV and scaling factor subfields in the QoS control field depending on actual QS and it also shows how the AP calculates STA QS.

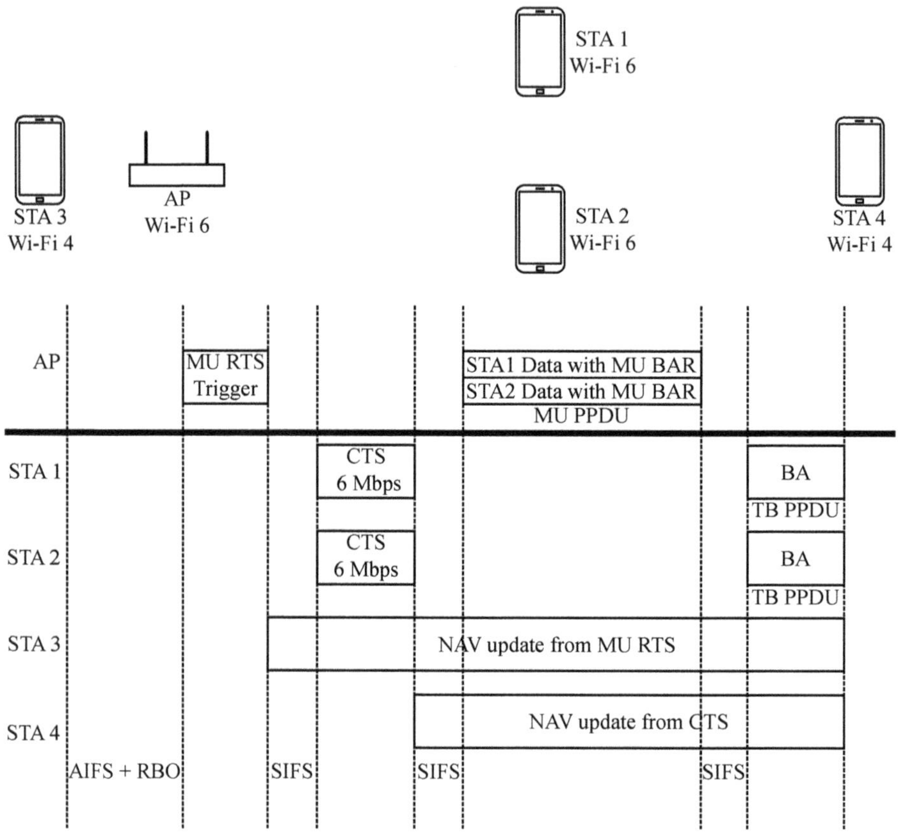


**Figure 3.12** Downlink MU with MU RTS and uplink MU for BAs.

$$SF = \begin{cases} 16, & \text{if scaling factor subfield} = 0 \\ 256, & \text{if scaling factor subfield} = 1 \\ 2048, & \text{if scaling factor subfield} = 2 \\ 32768, & \text{if scaling factor subfield} = 3 \end{cases} \tag{3.1}$$

The second method involves STA sending unsolicited BSR to the AP in the BSR control subfield of HT control field in QoS null, QoS data, or management frame. This method of reporting BSR in the BSR control subfield can be used by

**Table 3.9**
Queue Size Encoding in QoS Control Field

| *Actual Queue Size (QS) at STA in bytes* | *Unscaled Value (UV) Subfield* | *Scaling Factor Subfield* | *Scaling Factor (SF) in Bytes* | *Calculated STA QS at AP in Bytes* |
|---|---|---|---|---|
| 0 | 0 | 0 | 16 | $SF \cdot UV$ |
| $0 < QS \leq 1008$ | $\lceil \frac{QS}{16} \rceil$ | | | |
| $1008 < QS \leq 1024$ | 0 | 0 | 256 | 1024+ |
| $1024 < QS \leq 17152$ | $\lceil \frac{QS-1024}{256} \rceil$ | | | $SF \cdot UV$ |
| $17152 < QS \leq 17408$ | 0 | 2 | 2048 | 17408+ |
| $17408 < QS \leq 146432$ | $\lceil \frac{QS-17408}{2048} \rceil$ | | | $SF \cdot UV$ |
| $146432 < QS \leq 148480$ | 0 | 3 | 32768 | 148480+ |
| $148480 < QS \leq 2147328$ | $\lceil \frac{QS-148480}{32768} \rceil$ | | | $SF \cdot UV$ |
| $QS > 2147328$ | 62 | 3 | 32768 | $> 2147328$ |
| unknown | 63 | 3 | 32768 | unknown |

STAs only if the AP indicates support for it in the BSR support subfield of HE MAC capabilities information field in the HE capabilities IE. In this method, a single frame can contain the buffer status of multiple ACs, but this provides only a coarse summary view of QS across all ACs and does not provide precise QS information per TID. Since multiple TIDs map to one AC, the BSR control method despite being more efficient presents some challenges to the AP in per TID UL resource allocation. Figure 3.13 shows the format of BSR control subfield. The STA indicates the ACs for which it is reporting buffer status by setting 1 to the appropriate AC in the 4-bit AC indicator (ACI) bitmap subfield (one bit for each of the four ACs). The 2-bit ACI high subfield specifies a high-priority AC as determined by the STA. The queue size all subfield represents the total queue size summed across all ACs specified in the ACI bitmap, while the queue size high represents the QS of the high-priority AC specified in ACI high. Both queue size all and queue size high represent the quantized QS in units of $SF$ bytes, where $SF$ is derived from scaling factor subfield following (3.1). A value of 254 for queue size all or queue size high implies the amount of buffered traffic is greater than $254 \cdot SF$ bytes while a value of 255 implies unknown amount of buffered traffic. The delta TID subfield can be used to compute the number of TIDs for which there is buffered traffic by adding Delta TID to the number of bits that are set to 1 in ACI bitmap. There is one exception in the case when ACI bitmap is all zeros, which indicates that all 8 TIDs have buffered traffic. For example, if ACI bitmap = 0110 and delta TID = 1, then the number of TIDs with buffered traffic is 3.

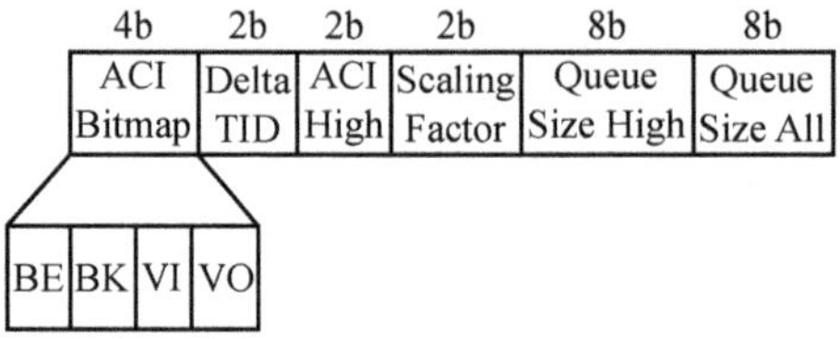


**Figure 3.13** BSR control subfield format.

The last method of obtaining BSR is by AP explicitly soliciting a BSR from multiple STAs concurrently using a buffer status report poll (BSRP) trigger frame. In response to BSRP, each of the polled STAs transmits a TB PPDU with one or more QoS null frames with BSR included as per the QoS control method or BSR control method (if AP supports it) as described earlier. A BSRP trigger frame can also be sent aggregated with other DL frames in a MU PPDU.

### 3.3.6 MU EDCA

In UL MU, the AP contends for the medium to send MU RTS or trigger frame but the transmit opportunity (TxOP) secured is utilized by STAs. Essentially, the AP is contending for airtime for not only the downlink traffic, but also for the uplink traffic from Wi-Fi 6 STAs. However, the chance for the AP to obtain a successful TXOP decreases severely as the number of associated stations increases. So, to improve the channel access probability of APs while lowering the channel access probability of Wi-Fi 6 STAs, Wi-Fi 6 introduces a new set of enhanced distributed channel access (EDCA) parameters called MU EDCA parameters that are advertised in the MU EDCA parameter set IE of beacon and probe frames along with the regular EDCA parameters in the WMM field. The MU EDCA parameters contain medium contention parameters such as AIFSN, $CW_{min}$, $CW_{max}$, and a timer parameter called MU EDCA timer per AC. The MU EDCA parameters must be followed for medium contention by all associated Wi-Fi 6 STAs that had the opportunity to transmit QoS data in a TB PPDU in the last MU EDCA timer $\times 8$ TUs (where 1 time unit (TU) = 1024 $\mu$s). If no trigger frame was received in the last MU EDCA timer $\times 8$ TUs, the STA can follow the regular EDCA parameters for medium contention. This MU EDCA mechanism allows for a fair share of medium between AP and participating STAs in UL MU while not affecting the behavior of previous generation Wi-Fi clients. The AP can change the values of MU EDCA parameters via beacon frame anytime and the associated Wi-Fi 6 STAs are required to follow the most recent MU EDCA parameters advertised by AP.

The MU EDCA feature can also be used to influence Wi-Fi 6 STAs that recently transmitted QoS data in a TB PPDU to wait a little longer for the next basic trigger frame by increasing the medium contention time for UL SU.

## 3.4 TARGET WAKE TIME

Wi-Fi 6 introduces a new power save mechanism called target wake time (TWT) that addresses some of the drawbacks in existing power save mechanisms. The existing PS schemes require a STA in power save state to wake up for every beacon, check the TIM IE for traffic presence, and retrieve buffered frames using PS-Poll or UAPSD, both of which have uncertain STA wake times. Moreover, during the wake time, the STA has to contend for the medium against other STAs. Although this works reasonably well for many use cases, it is clearly inefficient for known traffic patterns such as voice over Wi-Fi, which is a periodic bidirectional traffic every 20 ms. Another example could be an Internet of Things (IoT) sensor node that reports temperature or humidity every 1 hour. For such known traffic patterns, it is more efficient on power consumption if the STA and AP agreed upon a reserved, contention-free time period when the STA can wake up, transmit, or receive some traffic and go back to sleep. The TWT power save mechanism accomplishes this by enabling STAs to enter into a wake schedule agreement with the AP, so that STAs can be awake only when needed and remain in sleep otherwise. This also has an added benefit of reducing medium contention. Before describing TWT operation, let us define some common TWT terminologies. A TWT session period or service period (SP) is the time period when a STA is expected to be awake and can transmit or receive data. A TWT agreement between an AP and STA defines the parameters of the TWT SP. A TWT requester is a STA that requests for a TWT agreement and TWT responder is an AP that can accept, reject, or suggest alternate parameters for TWT agreement. To facilitate the exchange of TWT request and TWT response messages leading to a TWT agreement, a new TWT IE is defined in Wi-Fi 6. A TWT IE contains all the information required to negotiate different types of TWT agreements between AP and STAs and it can be present in beacon, probe response, reassociation request and response frames. Once a TWT agreement is reached, the STA can be in sleep until its allotted TWT SP begins and it can transmit or receive data within its assigned TWT SP. Although it is recommended that STAs transmit only within their assigned TWT SPs, the standard gives the option for STAs to transmit outside TWT SP as well by following EDCA. Depending on the traffic pattern, a STA can establish up to eight TWT agreements with an AP.

### 3.4.1 TWT Agreement Setup

The first step in TWT agreement setup or negotiation is initiated by TWT requester (STA) using a TWT request message to TWT responder (AP). The TWT requester can specify one of three TWT setup commands in a TWT request message as follows:

1. *Request*: Indicates the STA is requesting for a TWT agreement with parameters specified completely by the AP.
2. *Suggest*: The STA requests for a TWT agreement along with a suggested list of TWT parameters, but it is open to an alternate suggestion of parameters by the AP.
3. *Demand*: The STA requests for a TWT agreement with specified TWT parameters and would not accept any alternate suggestion from AP.

The AP in response can specify one of four TWT setup commands in a TWT response message as follows:

1. *Accept*: TWT agreement is setup using the parameters specified in TWT request message.
2. *Reject*: The AP rejects the TWT agreement request.
3. *Alternate*: AP suggests an alternate set of TWT parameters.
4. *Dictate*: AP proposes a final set of TWT parameters.

Depending on the TWT response from AP, the STA might need another round of TWT request, response frame exchanges to conclude the TWT agreement setup. An AP can have multiple TWT agreements with multiple STAs and their TWT SPs are allowed to overlap.

### 3.4.2 Types of TWT Agreements

There are two types of TWT agreements: individual and broadcast.

#### 3.4.2.1 Individual TWT Agreement

Individual TWT agreement is an agreement negotiated between an AP and an individual STA. The key parameters that define an individual TWT agreement are TWT, TWT wake interval, and minimum TWT wake duration. TWT is the time in $\mu$s when the TWT SP begins and the STA is expected to be awake. TWT wake interval denotes the time interval between consecutive TWT SPs if it is periodic and this can be multiple beacon intervals. Minimum TWT wake duration indicates the minimum time period the STA would stay awake since the beginning of TWT SP. There are different modes of TWT operation as listed below:

1. Implicit or explicit;
2. Announced or unannounced;
3. Trigger-enabled or nontrigger-enabled.

In explicit or aperiodic TWT operation mode, the TWT value is notified by a directed message from AP to STA before the next TWT session. In implicit or periodic TWT operation mode, the STA only knows the first TWT value and it calculates the subsequent TWT values using TWT interval. Inside a TWT SP, announced TWT operation mode requires the STA to send PS poll or UAPSD trigger frames to retrieve buffered frames from AP, whereas in unannounced mode, the AP sends data to STA assuming STA is awake without waiting for any prior frame from STA. In trigger-enabled TWT operation mode, the AP is required to schedule STA transmissions inside a TWT SP using a trigger frame whereas in nontrigger-enabled mode, STA is free to transmit anytime inside its TWT SP. Figure 3.14 illustrates an implicit, unannounced TWT operation using individual TWT agreement.

#### 3.4.2.2 Broadcast TWT Agreement

In a broadcast TWT agreement, the AP sets up a TWT SP with a group of STAs and advertises the TWT parameters in beacon frames for all existing broadcast TWT agreements. STAs can request an AP to either join an existing broadcast TWT agreement or create a new broadcast agreement. STAs participating in a broadcast

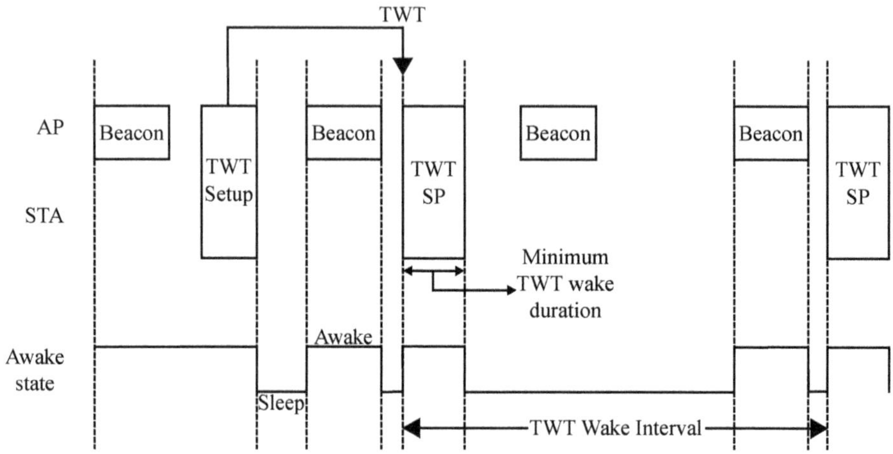


**Figure 3.14** Implicit TWT operation using individual TWT agreement.

TWT agreement are required to wake up and receive only those beacons containing information for their broadcast TWT SPs. The key parameters of a broadcast TWT agreement are next target beacon and listen interval. Next target beacon specifies transmission time of the next beacon frame containing TWT information corresponding to the broadcast TWT SP. Listen interval specifies the time interval between consecutive beacons containing information relevant for the broadcast TWT SP. The AP periodically updates the TWT parameters for broadcast TWT SP in beacon frame to enable member STAs to follow the schedule of broadcast TWT SP. Figure 3.15 illustrates broadcast TWT operation. The different TWT operation modes for individual TWT agreement apply to broadcast TWT agreement as well. One of the main benefits of broadcast TWT agreement is its simplicity and less management frame overhead compared to individual TWT agreement. Since a broadcast TWT agreement involves a group of STAs, it may be desirable for an AP to choose trigger-enabled TWT operation mode; otherwise the group of STAs will contend among themselves during their TWT SP.

### 3.4.3 TWT Summary

In summary, the AP has control over how to group TWT-enabled stations and allocate wake time periods, but the STA can influence the AP in several ways. TWT PS scheme provides more certainty to a STA on its wake time, maximizing sleep

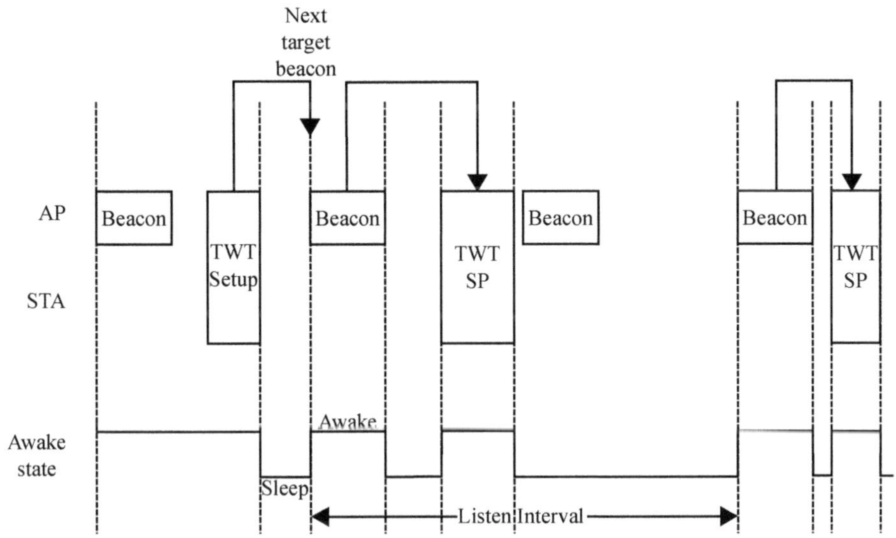


**Figure 3.15** TWT operation using broadcast TWT.

time while reducing time wasted on collisions. If all the STAs supported TWT, the AP can emulate time division multiple access (TDMA) style scheduling with fair airtime allocation, collision-free operation within a BSS, and allow significant power savings for the STA [4]. A STA can make dramatic power savings with TWT if it knows its traffic pattern, which is possible in applications such as IoT sensors and voice or video calling over Wi-Fi. Finally, it is the STA that has to initiate the TWT request and it is completely the STA's choice whether it wants to use TWT or one of the legacy PS schemes depending on the use case.

## 3.5 BSS COLORING

BSS coloring is a Wi-Fi 6 feature intended for dense AP deployments. It enables a Wi-Fi 6 device to classify a received HE PPDU as intra-BSS or inter-BSS. Here, intra-BSS refers to PPDUs belonging to the same BSS, while inter-BSS refers to PPDUs of an overlapping BSS (OBSS). This information can be used by APs and STAs in a dense AP deployment to ignore decoding certain PPDUs and to exploit concurrent transmit opportunities under special conditions. This reuse of a channel (or spectrum) at different locations to transmit concurrently while keeping the cochannel interference within acceptable limits is called the spatial reuse feature and BSS coloring provides the foundation for that.

With BSS coloring, every BSS is assigned a BSS color indicated by a number between 1 and 63. The implementation of BSS color assignment can be either distributed or centralized. In a distributed implementation, each AP in a BSS can pick its BSS color either independently or collaboratively with its cochannel neighbors. An example of a collaborative distributed implementation is an AP discovering its cochannel neighbors (over wired or wireless medium) and then picking a BSS color different from its cochannel neighbors. An example of an independent distributed implementation is one where each AP randomly picks its BSS color. A centralized implementation is applicable in a controller managed enterprise AP deployment, where the controller assigns the channel plan and BSS color for each AP.

Every AP advertises its support for BSS color feature and its current BSS color in the BSS color information field of HE operation IE in the beacon frame. In the case of multiple VAPs or BSSIDs cohosted in a physical AP, all the VAPs of a physical AP are assigned the same color. All HE PPDUs transmitted in a BSS by either the AP or its associated STAs are required to mark the BSS color in the HE-SIG-A field of PHY header and this value should match the advertised BSS color of AP. This enables a Wi-Fi 6 device to classify any received HE PPDU as intra-BSS or inter-BSS by looking at the PHY header.

### 3.5.1 BSS Color Collision

BSS color assignment may sometimes result in a BSS color collision wherein more than one cochannel physical AP advertises the same BSS color. For example, with independent random BSS color assignment following uniform distribution, the BSS color collision probability in a deployment having four cochannel APs within hearing range of each other is $1 - \frac{63 \cdot 62 \cdot 61 \cdot 60}{63^4} = 9.24\%$. A BSS color collision is detected by an AP or STA if it receives any HE PPDU with the same color as its BSS, but none of the three address fields in the MAC header match the BSSID for its associated AP or that of other cohosted VAPs. If a STA detects a BSS color collision, it notifies its associated AP of a color collision event using a BSS color collision event report frame.

### 3.5.2 BSS Color Change Announcement

An AP can change its BSS color either voluntarily or in response to a BSS color collision event. AP advertises a pending BSS color change using the BSS color change announcement element in the beacon, probe response, and association response frames. The BSS color change announcement contains the new BSS color information and a color switch countdown field that indicates the number of TBTTs pending before the color change happens. The pending time for color change is chosen to ensure all associated STAs have a chance to listen to at least one beacon frame. During this time, AP temporarily disables BSS coloring by setting the BSS color disabled subfield of BSS color information field to 1 in the HE operation IE but continues advertising the existing BSS color in the HE operation IE and marking the existing color in transmitted HE PPDUs. When it is time for the BSS color change to take effect, the AP sets the BSS color disabled subfield to 0 and starts advertising the new BSS color in HE operation IE. After that, the AP and its associated STAs are required to mark the new BSS color in HE PPDU transmissions.

## 3.6 SPATIAL REUSE

With the ability to detect inter-BSS PPDU using HE-SIG-A early on in PPDU reception, the spatial reuse (SR) feature enables devices to abort reception of certain inter-BSS PPDUs and transmit on top of them while keeping interference within limits. There are two different mechanisms for spatial reuse [5]: OBSS packet detect (OBSS PD)-based SR for TxOP from HE SU, HE ER SU, and HE MU PPDUs, and parameterized spatial reuse (PSR)-based SR for TxOP from HE TB PPDUs.

### 3.6.1 OBSS PD-based SR

As per carrier sense multiple access collision avoidance (CSMA-CA) protocol, a device is supposed to consider medium as busy if it senses any signal with power greater than -62 dBm or if it detects a Wi-Fi PHY header with power greater than -82 dBm. However, most Wi-Fi implementations detect packets even below -82 dBm and may consider the medium as busy. OBSS PD-based SR allows a Wi-Fi 6 device (AP or STA) to consider the medium as idle, and transmit if it receives an inter-BSS PPDU at a power level less than a chosen OBSS PD threshold. The TxOP so gained due to SR is referred as SR TxOP and is depicted in Figure 3.16. A device can chose an OBSS PD threshold value between OBSS $PD_{min}$ = -82 dBm and OBSS $PD_{max}$ = -62 dBm with the additional constraint that the Tx power in dBm during SR TxOP denoted by $TP_{sr}$ and OBSS PD threshold satisfy (3.2).

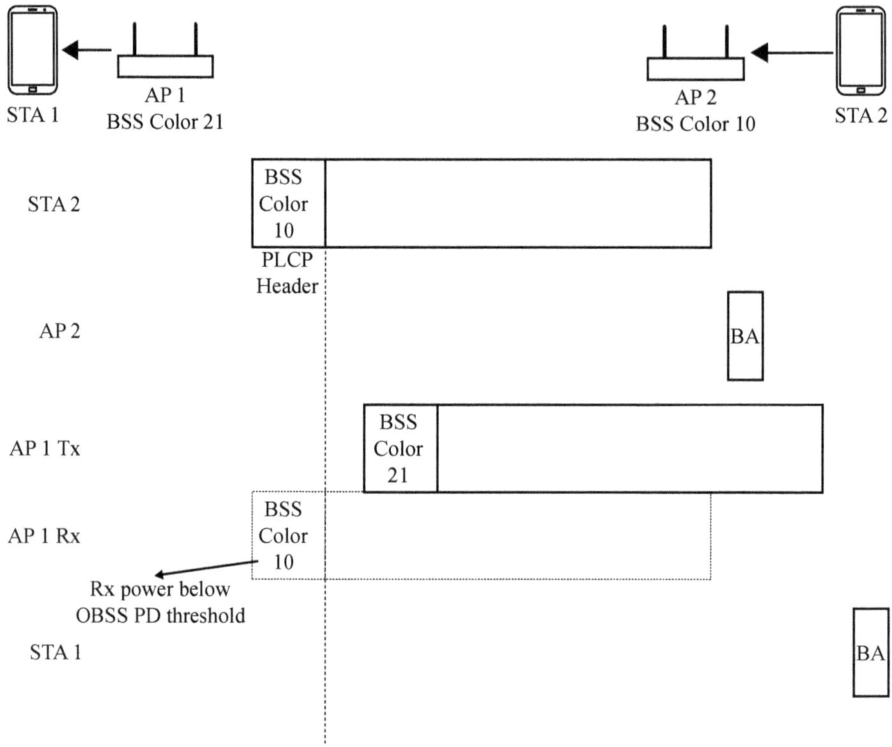


**Figure 3.16** Spatial reuse TxOP.

$$\text{OBSS PD} \leq \text{OBSS PD}_{min} + (TP_{ref} - TP_{sr}) \tag{3.2}$$

where,

$$TP_{ref} = \begin{cases} 21 \text{ dBm for STA,} \\ 21 \text{ dBm for AP with Tx streams capability} < 3 \\ 25 \text{ dBm for AP with Tx streams capability} > 2 \end{cases} \tag{3.3}$$

Equation (3.2) ensures that the interference caused by SR is within acceptable limits. Figure 3.17 illustrates the allowed range of OBSS PD and Tx power during SR TxOP. For example, if a two stream capable AP wants to transmit at 15 dBm during SR TxOP then it has to ensure that its OBSS PD threshold is upper bounded by $-82 + (21 - 15)\text{dBm} = -76$ dBm. Similarly, if a STA applies an OBSS PD threshold of $-72$ dBm, then its $TP_{sr}$ must be upper bounded by $21 - 82 + 72\text{dBm} =$ 11 dBm. If necessary, the AP can also advertise a OBSS $PD_{max}$ value different from the default -62 dBm using the non-SRG OBSS $PD_{max}$ field in spatial reuse parameter set IE in beacons and association response frames and associated STAs would follow them.

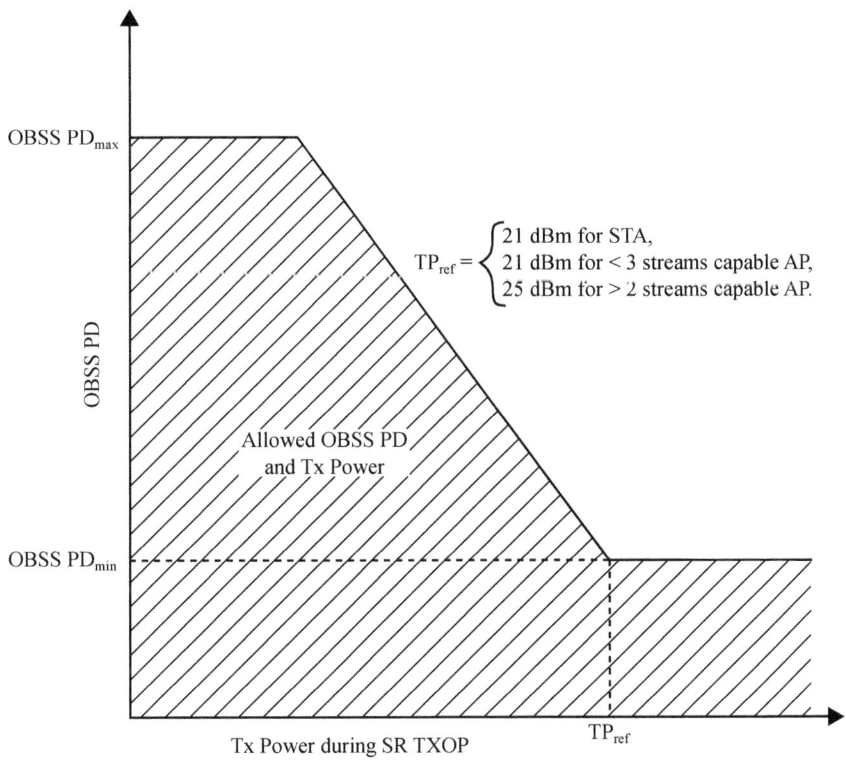


**Figure 3.17** OBSS PD and Tx power during SR TxOP.

#### 3.6.1.1 Dynamic OBSS PD with Transmit Power Adjustment

To increase the number of opportunities for SR TxOP, devices can also implement dynamic OBSS PD with Tx power adjustment instead of a static OBSS PD threshold. This method enables SR transmission on top of all inter-BSS HE PPDUs received with power less than OBSS $PD_{max}$. Here, the OBSS PD threshold is set equal to MIN(OBSS $PD_{max}$, $RP_{inter\text{-}BSS}$) where, $RP_{inter\text{-}BSS}$ is the received power of inter-BSS HE PPDU. The Tx power during SR TxOP $TP_{sr}$ is dynamically adjusted to be below OBSS $PD_{min} - RP_{inter\text{-}BSS} + TP_{ref}$, thereby satisfying (3.2).

#### 3.6.1.2 Dual NAV Operation

As SR allows aborting an inter-BSS PPDU reception, it is important to understand how NAV updates are done for inter-BSS PPDUs. A device using the SR feature is required to maintain two NAVs: intra-BSS NAV and inter-BSS NAV. The intra-BSS NAV is updated on reception of any intra-BSS PPDU, while the inter-BSS NAV is updated upon reception of an inter-BSS PPDU that doesn't qualify for SR. Essentially, NAV updates are exempted for inter-BSS PPDUs with power below the OBSS PD threshold. The medium is considered busy if either intra-BSS NAV or inter-BSS NAV is nonzero.

#### 3.6.1.3 SRG OBSS PD

Sometimes it may be preferable to use different OBSS PD thresholds depending on the BSS color of incoming HE PPDU. For example, an AP may want to use a different set of OBSS $PD_{min}$, OBSS $PD_{max}$ values for HE PPDUs received from its closest cochannel neighbor while using different values for far away neighbors or neighbors of a different network. The spatial reuse group (SRG) OBSS PD mechanism allows to define a group of BSS colors for which a custom set of OBSS PD values called SRG OBSS $PD_{min}$, SRG OBSS $PD_{max}$ are applied, while BSS colors that are not part of SRG group follow the non-SRG OBSS PD or default OBSS PD values. The AP advertises the BSS colors that are part of SRG group and the associated SRG OBSS $PD_{min}$, SRG OBSS $PD_{max}$ values in the spatial reuse parameter set IE.

### 3.6.2 PSR-based SR

The PSR-based SR mechanism permits SR on top of TB PPDUs. In this mechanism, the AP specifies the parameters for PSR-based SR in the UL spatial reuse subfield

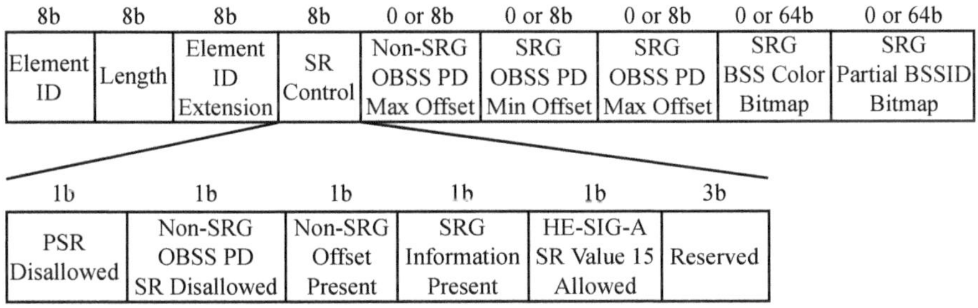


**Figure 3.18** Spatial reuse parameter set IE.

of common info field in every trigger frame. The UL spatial reuse subfield indicates whether PSR-based SR is allowed and specifies a $PSR_{input}$ parameter that constrains the maximum Tx power for performing SR on top of the TB PPDU that immediately follows the trigger frame. The $PSR_{input}$ parameter is computed by the AP following (3.4), where $TP_{trigger}$ is the Tx power of the trigger frame and $I_{AP}$ is the maximum interference power that the AP can handle without affecting TB PPDU reception.

$$PSR_{input} = TP_{trigger} + I_{AP} \tag{3.4}$$

Any Wi-Fi 6 device receiving an inter-BSS trigger frame at a power level $RP_{trigger}$ is allowed to do PSR-based SR transmission during the following TB PPDU provided its Tx power is below $PSR_{input} - RP_{trigger}$. By following this Tx power constraint, it can be verified that the interference power at the AP during TB PPDU reception will be below $I_{AP}$. The PSR-based SR mechanism provides greater flexibility for the AP to control the $PSR_{input}$ parameter on a per trigger frame basis and even disallow SR for certain TB PPDUs if required.

### 3.6.3 Practical Challenges in SR

One of the challenges with SR is that the transmitter cannot selectively protect certain HE PPDUs from SR by other devices. Although PSR-based SR allows such control on a per trigger frame basis, it is available only for the UL TB PPDUs. Another challenge is the difficulty in estimating the optimal MCS when performing SR transmission with reduced Tx power.

## 3.7 ENHANCED MULTI-BSSID ADVERTISEMENT

Wi-Fi 6 introduces a new mode of capability advertisement called EMA mode applicable to multiple VAP operation. This feature significantly reduces the management frame overhead in dense AP deployments configured with multiple VAPs. The regular cohosted mode of multiple VAP operation, illustrated in Figure 3.19, requires each AP to send a beacon frame per VAP every TBTT period and each VAP is required to send a probe response frame in response to a broadcast or wildcard probe request frame. This adds management frame overhead in a dense AP deployment, where multiple cochannel APs may be within hearing range of each other. In public venues with a high client count such as stadiums or train stations, this results in a significant fraction of airtime being spent on beacons and probe responses.

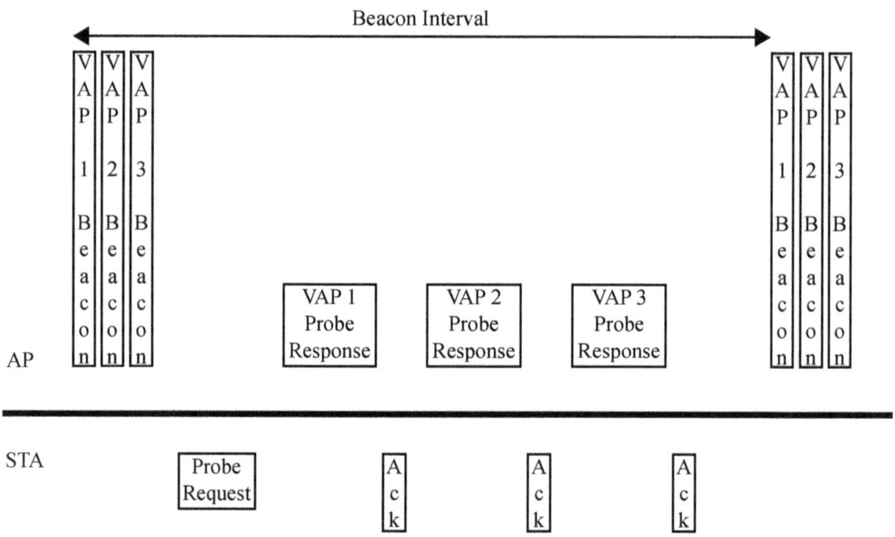


**Figure 3.19** Cohosted mode of multiple VAP operation.

The EMA mode allows an AP to assign one or more BSSIDs as transmitted BSSIDs and the remaining are defined as nontransmitted BSSIDs. In the EMA mode, beacons and probe responses are transmitted only for the transmitted BSSIDs. The capability information of all nontransmitted BSSIDs is aggregated and packaged inside a multiple BSSID IE that is piggybacked on a transmitted BSSID's beacon and probe response frame. The multiple BSSID IE contains one or more nontransmitted BSSID profile subelements corresponding to each nontransmitted BSSID. A nontransmitted BSSID profile subelement contains the BSSID index, SSID, and capability information for a nontransmitted BSSID. The BSSID Index is an identifier for each nontransmitted BSSID and can take on any value between 1 and $2^{n-1}$, where $2^n$ is the maximum number of BSSIDs in the multiple BSSID set. A BSSID index of 0 refers to the reference BSSID that is assigned as one of the transmitted BSSIDs. To reduce overhead, the capability elements of a nontransmitted BSSID that match the same values as that of the transmitted BSSID are skipped in the nontransmitted BSSID profile subelement and this is called inheritance of element values. In the EMA mode, all the BSSIDs in a multiple BSSID set are required to follow a specific style of MAC address assignment in which all of them share the first $48 - n$ bits of MAC address. This allows the MAC address of a nontransmitted BSSID to be derived from the BSSID Index. For example, if there are $2^4 = 16$ BSSIDs in a multiple BSSID set, then all the 16 BSSIDs must share the first $48 - 4$ bits. Note that the AID space is shared by all BSS, and the TIM IE indicates the presence or absence of traffic to all STAs associated to either the transmitted or nontransmitted BSS. All Wi-Fi 6 STAs are required to have the capability to discover nontransmitted BSS by parsing the multiple BSSID IE in beacons and probe responses. An AP indicates support for EMA mode using the EMA support field in extended capabilities IE. Figure 3.20 illustrates the EMA mode of multiple VAP operation.

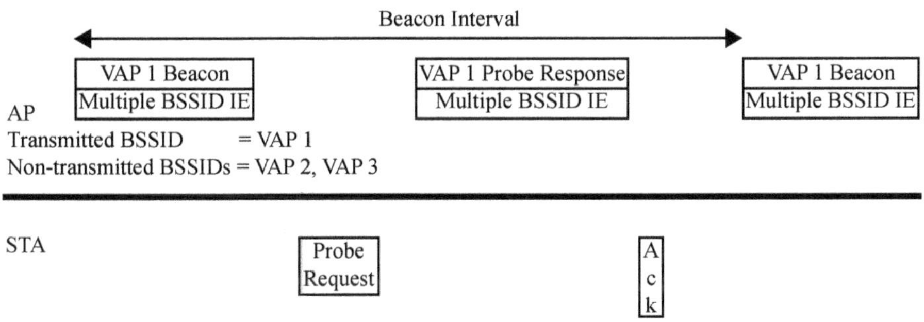


**Figure 3.20** EMA mode of multiple VAP operation.

### 3.7.1 Profile Periodicity

Sometimes, it is not possible to fit all the nontransmitted BSSID profiles in a single beacon frame especially when there are a large number of nontransmitted BSSIDs since some implementations have limitations on the maximum beacon frame size. In such cases, each beacon frame contains only a partial list of nontransmitted BSSID profiles and the information of all nontransmitted BSSID profiles is spread over multiple beacons separated by beacon interval time duration. The minimal number of consecutive beacon intervals required to discover all nontransmitted BSSID profiles is called profile periodicity. The full set Rx periodicity field of multiple BSSID configuration IE indicates the profile periodicity of an AP. The DTIM period for all nontransmitted BSSIDs is set to be a multiple of profile periodicity so that STAs connected to a nontransmitted BSSID always get their profile information in their respective DTIM beacon.

## 3.8 PREAMBLE PUNCTURING

One of the practical problems with wide BW modes (especially 80 MHz or 160 MHz) is the limited channel availability owing to the potential interference from other legacy APs in the network overlapping either on the primary subchannel or secondary subchannels. One way to mitigate this problem is by doing dynamic channel bonding or dynamic BW adaptation [6], wherein a 160-MHz device (AP or STA) would perform clear channel assessment (CCA) on every 20 MHz subchannel and adapt the transmit BW on a per PPDU basis as follows:

- 160 MHz if all subchannels are idle;
- 80 MHz if secondary 80-MHz subchannel is busy;
- 40 MHz if secondary 40-MHz subchannel is busy;
- 20 MHz if secondary 20-MHz subchannel is busy.

The dynamic channel bonding technique can be applied in both Wi-Fi 5 and Wi-Fi 6 devices. An alternative but partial solution to this problem is preamble puncturing in Wi-Fi 6, which offers a way to do noncontiguous channel bonding for DL traffic. Preamble puncturing takes advantage of DL MU-OFDMA to avoid transmission in the busy secondary subchannels by not allocating RUs in those positions. In addition, the legacy portion of the HE MU PPDU PHY header, which is usually duplicated on every subchannel is punctured on the busy secondary subchannels. Hence, with preamble puncturing APs can transmit in 80- or 160-MHz BW to Wi-Fi 6 clients supporting preamble punctured PPDU reception while not causing interference to legacy devices that overlap on the secondary subchannels. The punctured preamble Rx subfield of HE PHY capabilities information field in the HE capabilities IE indicates support for preamble punctured PPDU reception. One limitation is that preamble puncturing is not possible on the primary subchannel. Figure 3.21 compares dynamic channel bonding and preamble puncturing options in 80-MHz BW. Clearly, preamble puncturing makes better use of the spectrum as compared to dynamic channel bonding.

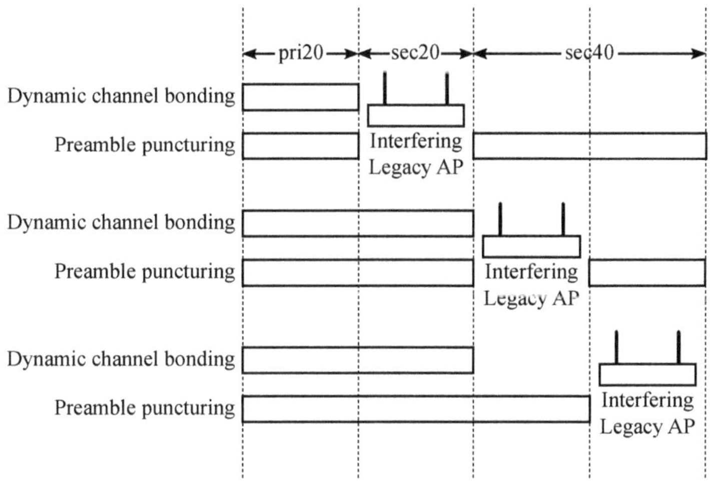


**Figure 3.21** Dynamic channel bonding and preamble puncturing in 80-MHz BW.

## 3.9 OPERATING MODE INDICATION

One of the long-standing problems in Wi-Fi is that associated STAs cannot change their BW or spatial stream capability without reassociating with an AP. However, a reassociation event causes traffic disruption, which is undesirable in several use cases. There are some practical scenarios where a STA may want to temporarily disable a supported feature post association. One such example scenario is relevant in consumer electronic products wherein a client device temperature may exceed acceptable limits owing to its small physical form factor and simultaneous activity in multiple blocks, such as display, processor, speaker, or wireless, along with high ambient temperature. In such a scenario, the client device can get thermal relief by temporarily switching to a lower BW and spatial stream for both receive and transmit. A useful feature in Wi-Fi 6 that addresses this problem is operating mode indication (OMI).

OMI provides a mechanism for an associated STA to change its receive or transmit operating mode without reassociating with the AP. This notification of operation mode change is done using a newly defined operating mode (OM) control subfield in the HT control field of any unicast QoS data, QoS null, or management frame. The AP updates its view of STA capabilities upon reception of OM control subfield and acknowledges it with an ACK frame. The STA is permitted to change its operating mode as indicated in the OM control subfield upon receiving acknowledgment from the AP. Figure 3.22 shows the OM control subfield format and Table 3.10 describes the usage for different transmit and receive operating mode indicators.

| 3b | 2b | 1b | 3b | 1b | 1b | 1b |
|---|---|---|---|---|---|---|
| Rx NSS | Channel Width | UL MU Disable | Tx NSTS | ER SU Disable | DL MU-MIMO Resound | UL MU Data Disable |

**Figure 3.22** Operating mode control subfield format.

## 3.10 HE CAPABILITY ADVERTISEMENT

The supported HE features of an AP or STA are specified by the HE capabilities IE, HE operation IE and extended capabilities IE. Note that some of the mandatory HE features do not have any capability advertisement fields as all HE devices are assumed to support them. For example, there are no fields that advertise support for OFDMA. Table 3.11 lists some of the common HE capability information advertised in beacons and probes.

## 3.11 WHEN TO DO SU, MU-MIMO AND MU-OFDMA

One of the challenges in implementing a Wi-Fi 6 scheduler is choosing whether to use SU, MU-MIMO, or MU-OFDMA since there are many scenarios where MU-MIMO and MU-OFDMA perform worse than SU. There are several requirements

**Table 3.10**
Transmit and Receive Operating Mode Indicators in OM Control Subfield

| *Subfield* | *Usage* |
|---|---|
| Rx NSS | Indicates maximum receive NSS capability of STA. |
| Channel Width | Indicates maximum receive and transmit BW capability of STA. |
| ER SU Disable | Indicates the STA support for ER SU PPDU reception. |
| DL MU-MIMO Resound | Used by STA to request AP to redo MU sounding. This can be useful in scenarios where the STA becomes aware that the CV is stale. |
| Tx NSTS | Indicates the maximum NSS that the STA can support in a HE TB PPDU transmission. |
| UL MU Disable | Indicates STA support for UL MU. This is used by a STA in scenarios when it wants to contend for the medium and have greater control on its transmission schedule. |
| UL MU Data Disable | If set to 1, it indicates that the STA cannot support responses to a basic trigger frame, but can support responses to other type trigger frame types such as MU BAR, BSRP. |

and challenges for DL MU-MIMO: good channel quality feedback, which is a problem in low SNR; frequent MU sounding to avoid stale CV; similar payload size and MCS across clients involved in a MU PPDU for optimal airtime usage; and MU transmission often has to use lower MCS compared to SU transmission due to SNR loss associated with null-steering. On the other hand, UL MU doesn't require sounding so it has less overhead, but requires proper UL transmit power control and UL rate adaptation to ensure TB PPDUs arrive at the appropriate power level and MCS at the AP. UL transmit power control may sometimes cause a reduction in MCS compared to SU depending on client location and grouping of clients. Stringent constraints on RU position for MU-OFDMA can result in suboptimal RU allocation that is mismatched with payload size distribution across clients. Moreover, OFDMA PHY rates are always lower than SU due to fewer number of data subcarriers. Hence, it is important for the scheduler to consider the various overheads and inefficiencies of MU compared to SU and the payload size and MCS distribution across clients before deciding among SU, MU-MIMO, and MU-OFDMA. Since this is a complex problem with no easy solution in general, the following worked-out examples calculate the best strategy for some specific DL traffic scenarios. This exercise can help shed some light onto the intricacies of the problem and provide some insights towards a possible solution.

**Table 3.11**

Common HE Capability Information

| *Information Element* | *Field Name* | *Subfield Name* | *Description* |
|---|---|---|---|
| VHT operation | VHT operation information | | Defines channel width and channel center frequency of the BSS. |
| HE capabilities | HE PHY capabilities information | Supported channel width set | Supported bandwidths. |
| HE capabilities | Supported HE-MCS and NSS set | | Supported list of HE MCS and NSS. |
| HE capabilities | HE PHY capabilities information | SU Beamformer / Beamformee | SU beamformer / beamformee capability. |
| HE capabilities | HE PHY capabilities information | MU Beamformer | MU beamformer capability. |
| HE capabilities | HE PHY capabilities information | Beamformee STS | Maximum NSS that the STA is capable of receiving in HE NDP frame minus 1. |
| HE capabilities | HE PHY capabilities information | Punctured preamble Rx | Punctured preamble Rx capability. |
| HE capabilities | HE PHY capabilities information | Full bandwidth UL MU-MIMO | Indicates support for UL MU-MIMO. |
| HE capabilities | HE PHY capabilities information | PSR-based SR support | Indicates support for PSR-based SR. |
| HE capabilities | HE MAC capabilities information | TWT requester support | TWT requester capability of STA. |
| HE capabilities | HE MAC capabilities information | TRS support | STA Rx support for frames with TRS control subfield. |
| HE capabilities | HE MAC capabilities information | BSR support | Indicates support for frames with BSR control subfield. |
| HE capabilities | HE MAC capabilities information | OM control support | Indicates Rx support for frames with OM control subfield. |
| HE capabilities | HE MAC capabilities information | PSR responder | Indicates Rx support for PSR-based SR transmission. |
| HE operation | BSS color information | | BSS coloring status. |
| Extended capabilities | Extended capabilities | EMA support | AP's EMA capability. |

### 3.11.1 Examples

To simplify the analysis, let us assume zero packet errors and collisions in the following examples.

#### 3.11.1.1 Example 1

Four 2 × 2 Wi-Fi 6 STAs are associated to a 4 × 4 Wi-Fi 6 AP in HE80 mode in the 5-GHz band. Let us assume that the DL SU PHY rate parameters for all STAs is identical at 2 spatial stream, MCS 7, 0.8 $\mu$s GI, 2× LTF. Let the buffered DL traffic to each STA be 20000 bytes of BE access category including all MAC headers, MSDU and MPDU delimiters. Let us calculate the airtime consumed to transmit this DL traffic to all four STAs in SU, MU-MIMO, and MU-OFDMA modes.

SU mode:

In the SU case, it is sufficient to calculate the airtime consumed for one STA and multiply the result by 4. Before each PPDU transmission, the AP needs to wait for a fixed time period of SIFS+3 slots and a random time period averaging 7.5 time slots for BE access category.

$$\text{Average airtime spent on backoff per STA} = 16 + 3 \cdot 9 + 7.5 \cdot 9\,\mu\text{s} \tag{3.5}$$
$$= 110.5\,\mu\text{s} \tag{3.6}$$
$$\text{Average airtime spent on backoff for 4 STAs} = 442\,\mu\text{s} \tag{3.7}$$

The SU PPDU PHY rate for 2 spatial stream, MCS 7, 0.8-$\mu$s GI is 720.58 Mbps. The SU PPDU airtime is composed of a PHY header portion and a payload portion. The airtime for the payload portion can be approximated by dividing the PSDU size by the PHY rate and then rounding up the result to the nearest multiple of 12.8+0.8 $\mu$s, which is the OFDM symbol duration including GI.

$$\text{SU PPDU airtime per STA} = (36 + 2 \cdot 6.4) + \left(\frac{20000 \cdot 8}{720.58}\right)\,\mu\text{s} \tag{3.8}$$
$$\approx 293.6\,\mu\text{s} \tag{3.9}$$
$$\text{Total airtime spent on data Tx to 4 STAs} = 1174.4\,\mu\text{s} \tag{3.10}$$

The BA from each STA is approximately 32 bytes (assuming 64-bit BA), which is transmitted at 24-Mbps legacy PHY rate after SIFS time.

$$\text{Airtime spent on getting ACK per STA} = \text{SIFS} + \left(20 + \frac{32 \cdot 8}{24}\right)\,\mu\text{s} \tag{3.11}$$

$$= 46.66\,\mu\text{s} \tag{3.12}$$

$$\text{Total airtime spent on getting ACKs} = 186.67\,\mu\text{s} \tag{3.13}$$

$$\text{Total airtime in SU mode} = 442 + 1174.4 + 186.67\,\mu\text{s} \tag{3.14}$$

$$= 1803.07\,\mu\text{s} \tag{3.15}$$

MU-MIMO mode:

In the DL MU-MIMO case, the total DL traffic to four STAs can be scheduled as two DL MU-MIMO sequences with each MU sequence serving a group of two STAs. The limitation of two STAs is because the AP is capable of transmitting a maximum of only four spatial streams. The BAs from STAs can be retrieved using UL MU-MIMO.

Since two DL MU-MIMO sequences are required to serve four STAs, the airtime spent on backoff for data transmission to four STAs is $2 \cdot 110.5 = 221\,\mu$s. The airtime for the payload portion of MU PPDU can be calculated by finding the maximum airtime across the two users. Let us assume the data rate for each user in the MU PPDU is 2 spatial stream MCS 5 which is two MCS lower than SU MCS owing to SNR loss from null-steering. The PHY rate for 2 spatial stream MCS 5 is 576.4 Mbps. Due to equal traffic and equal spatial stream allocation, it is sufficient to calculate airtime for one user, which would be $\frac{20000 \cdot 8}{576.4} \approx 285.6\,\mu$s when rounded up to an integer multiple of OFDM symbols. The PHY header portion of MU PPDU includes a variable HE-SIG-B field to specify the RU allocation or spatial stream allocation. The number of bits in HE-SIG-B field for MU-MIMO is approximately $21 \cdot (\text{Number of users}) + 10 \cdot \lceil \frac{\text{Number of users}}{2} \rceil$ bits, which works out to 52 bits. So the HE-SIG-B duration is $\frac{52}{6} \approx 12\,\mu$s.

$$\text{MU PPDU airtime} = 36 + 12 + 4 \cdot 6.4 + \frac{20000 \cdot 8}{576.4}\,\mu\text{s} \tag{3.16}$$

$$\approx 359.2\,\mu\text{s} \tag{3.17}$$

$$\text{Total airtime spent on data Tx} = 2 \cdot 359.2\,\mu\text{s} \tag{3.18}$$

$$= 718.4\,\mu\text{s} \tag{3.19}$$

Each MU PPDU is followed in SIFS time interval by a MU BAR trigger frame at 24-Mbps legacy PHY rate to solicit BAs using UL MU-MIMO. The number of bytes in a MU BAR trigger frame is approximately $28+9\cdot(\text{Number of users})$ bytes.

$$\text{Duration of MU BAR trigger frame} = 20 + \frac{(28+9\cdot 2)\cdot 8}{24}\,\mu\text{s} \quad (3.20)$$
$$\approx 36\,\mu\text{s} \quad (3.21)$$

To calculate the airtime for TB PPDU, let us assume UL BW of 80 MHz and that each user is allocated 1 spatial stream MCS 3 with 3.2$\mu$s GI, 4$\times$ LTF, which results in a PHY rate of 122.5 Mbps per user.

$$\text{Duration of TB PPDU containing BAs} = 40 + 2\cdot 12.8 + \frac{32\cdot 8}{122.5}\,\mu\text{s} \quad (3.22)$$
$$\approx 81.6\,\mu\text{s} \quad (3.23)$$
$$\text{Total airtime spent on getting ACKs} = 2\cdot(16+36+16+81.6)\,\mu\text{s} \quad (3.24)$$
$$= 299.2\,\mu\text{s} \quad (3.25)$$

Just before each DL MU-MIMO sequence, the AP performs a HE TB sounding sequence that starts with NDPA, NDP frames followed by a BFRP trigger frame to retrieve CBF reports using UL MU-MIMO. The NDPA contains $21 + 4\cdot(\text{Number of users})$ bytes while the BFRP contains $28+6\cdot(\text{Number of users})$ bytes. Let us assume that both NDPA and BFRP are transmitted at 24-Mbps legacy PHY rate.

$$\text{Duration of NDPA frame} = 20 + \frac{(21+4\cdot 2)\cdot 8}{24}\,\mu\text{s} \quad (3.26)$$
$$\approx 32\,\mu\text{s} \quad (3.27)$$
$$\text{Duration of NDP frame} = 36 + 4\cdot 6.4\,\mu\text{s} \quad (3.28)$$
$$= 61.6\,\mu\text{s} \quad (3.29)$$
$$\text{Duration of BFRP trigger frame} = 20 + \frac{(28+6\cdot 2)\cdot 8}{24}\,\mu\text{s} \quad (3.30)$$
$$\approx 36\,\mu\text{s} \quad (3.31)$$

The size of CBF report from each STA is given by $8\cdot \text{N}_\text{c} + (\text{N}_\text{s}\cdot \text{N}_\text{a}\cdot \text{bit precision})$ bits where, $\text{N}_\text{s}$ is the number of subcarriers for which feedback is reported and this depends on the BW and subcarrier grouping parameter $N_g$. Also $N_a$ is the number of angles per subcarrier that depends on $N_r$, $N_c$. For this example, $N_r = 4$, $N_c = 2$, which based on Table 2.7 results in $N_a = 10$. Let us assume 6-bit precision for the reported angles and $N_g = 4$, which based on Table 3.4 results in $N_s = 250$ for 80-MHz BW. The size of CBF report is therefore $8 \cdot 2 + (250 \cdot 10 \cdot 6) = 15016$ bits. Let us allocate 1 spatial stream MCS 3 with 3.2-$\mu$s GI, 4× LTF to each user for sending the CBF report in a TB PPDU using UL MU-MIMO.

$$\text{Duration of TB PPDU containing CBF reports} = 40 + 2 \cdot 12.8 + \frac{15016}{122.5}\,\mu\text{s} \tag{3.32}$$

$$\approx 193.6\,\mu\text{s} \tag{3.33}$$

So the total airtime consumed for one HE TB sounding sequence to two STAs is $110.5 + 32 + 16 + 61.6 + 16 + 36 + 16 + 193.6 = 481.7\,\mu$s.

$$\text{Sounding airtime for 4 STAs} = 2 \cdot 481.7\,\mu\text{s} \tag{3.34}$$

$$= 963.4\,\mu\text{s} \tag{3.35}$$

$$\text{Total airtime in MU-MIMO mode} = 963.4 + 221 + 718.4 + 299.2\,\mu\text{s} \tag{3.36}$$

$$= 2202\,\mu\text{s} \tag{3.37}$$

MU-OFDMA mode:

In the DL MU-OFDMA case, the traffic to all four STAs can be transmitted in a single MU PPDU and the BAs can be retrieved using UL MU-OFDMA.

$$\text{Average airtime spent on backoff for all STAs} = 110.5\,\mu\text{s} \tag{3.38}$$

Let us assume an RU allocation of one 242-tone RU per STA that fills all RUs except the center 26-tone RU in 80 MHz. Since the center 26-tone RU is impacted by DC interference, let us leave it unused. The PHY rate for 2 spatial stream, MCS 7, 242-tone RU with 0.8-$\mu$s GI is 172.06 Mbps. The airtime for the payload portion of MU PPDU is $\frac{20000 \cdot 8}{172.06} \approx 938.4\,\mu$s. The HE-SIG-B field for MU-OFDMA approximately has $21 \cdot (\text{Number of users}) + 10 \cdot \lceil \frac{\text{Number of users}}{2} \rceil + 18 \cdot (\text{Number of 20-MHz subchannels})$ bits, which works out to 176 bits. So the duration of HE-SIG-B is $\frac{156}{6} \approx 28\,\mu$s.

$$\text{Total airtime spent on data Tx} = 36 + 28 + 2 \cdot 6.4 + 938.4\,\mu\text{s} \tag{3.39}$$

$$= 1015.2\,\mu\text{s} \tag{3.40}$$

The MU PPDU is followed by a MU BAR trigger frame at 24-Mbps legacy PHY rate in SIFS time interval to solicit BAs from all four STAs using UL OFDMA. Let us assume a 242-tone RU allocation per STA and a data rate assignment of 1 spatial stream MCS 3, 3.2-$\mu$s GI, 4$\times$ LTF per STA for the TB PPDU, which results in data rate of 29.3 Mbps per STA.

$$\text{Duration of MU BAR trigger frame} = 20 + \frac{(28 + 9 \cdot 4) \cdot 8}{24}\,\mu\text{s} \tag{3.41}$$

$$\approx 44\,\mu\text{s} \tag{3.42}$$

$$\text{Duration of TB PPDU containing BAs} = 40 + 12.8 + \frac{32 \cdot 8}{29.3}\,\mu\text{s} \tag{3.43}$$

$$\approx 68.8\,\mu\text{s} \tag{3.44}$$

$$\text{Total airtime spent on getting ACKs} = 16 + 44 + 16 + 68.8\,\mu\text{s} \tag{3.45}$$

$$= 144.8\,\mu\text{s} \tag{3.46}$$

$$\text{Total airtime in MU-OFDMA mode} = 110.5 + 1015.2 + 144.8\,\mu\text{s} \tag{3.47}$$

$$= 1270.5\,\mu\text{s} \tag{3.48}$$

To summarize from (3.15), (3.37), and (3.48), it requires 1803 $\mu$s in SU mode, 2202 $\mu$s in MU-MIMO mode and 1271 $\mu$s in MU-OFDMA mode to send 20,000 bytes of downlink traffic to the four STAs. In terms of MAC layer throughput, SU, MU-MIMO and MU-OFDMA modes achieved 88.74 Mbps, 72.66 Mbps, and 125.88 Mbps, respectively. For this specific example scenario, MU-OFDMA is the best strategy as it consumes the least airtime of 1271$\mu$s, which is approximately 30% lower than SU mode. If the amount of DL traffic is lower than 20,000 bytes, it is clear that the MU-OFDMA benefit will be more than 30%. Observing the breakdown of airtime, it can be inferred that MU-OFDMA mainly saved airtime on backoff and retrieving ACKs. Both SU and MU-MIMO turned out to be inefficient primarily due to the small payload considering the high SU data rate of 720.58 Mbps. The MU-MIMO sounding overhead could have been reduced with a higher data rate assignment for the TB PPDU containing CBF reports, but in practice it is a trade-off against potential retransmissions for packet errors.

While it is commonly expected that MU-OFDMA is a good option for small payloads, the term "small payload" is relative and vague. In the context of this example, does 20,000 bytes fall in the small payload or large payload category? This worked-out example answers this question and highlights the importance of viewing all metrics in terms of airtime instead of bytes. In particular, this example also shows that the MSDU and MPDU size in bytes do not directly determine the optimal decision boundary between SU, MU-MIMO, and MU-OFDMA strategies as there are several other factors involved, such as number of MSDUs/MPDUs, BW, MCS, spatial streams, RU allocation, number of STAs, and so on.

#### 3.11.1.2 Example 2

Eight $2 \times 2$ Wi-Fi 6 STAs are connected to a $4 \times 4$ AP in HE20 mode. Let us assume the DL SU PHY rate to each STA is 2 spatial stream, MCS 9, 0.8-$\mu$s GI, $2\times$ LTF and the buffered DL traffic to each STA is 80,000 bytes of BE access category including MAC headers, MSDU and MPDU delimiters.

SU mode:

In SU mode, the traffic will be scheduled as 8 SU PPDUs to each STA. Refer to Example 3.11.1.1 for a detailed worked-out example.

$$\begin{aligned}\text{Total airtime spent on backoff} &= 8 \cdot 110.5\,\mu\text{s} && (3.49)\\ &= 884\,\mu\text{s} && (3.50)\end{aligned}$$

$$\begin{aligned}\text{Total airtime spent on data Tx} &= 8 \cdot \left(36 + 2 \cdot 6.4 + \frac{80000 \cdot 8}{229.4}\right)\mu\text{s} && (3.51)\\ &\approx 8 \cdot 2850.4\,\mu\text{s} && (3.52)\\ &= 22803.2\,\mu\text{s} && (3.53)\\ \text{Total airtime spent on getting ACKs} &= 8 \cdot 46.66\,\mu\text{s} && (3.54)\\ &= 373.28\,\mu\text{s} && (3.55)\\ \text{Total airtime in SU mode} &= 884 + 22803.2 + 373.28\,\mu\text{s} && (3.56)\\ &= 24060.48\,\mu\text{s} && (3.57)\end{aligned}$$

MU-MIMO mode:

If the same DL traffic is sent in MU-MIMO mode, it can be scheduled as four DL MU-MIMO sequences with each MU PPDU serving two STAs. Let us assume the DL MU-MIMO data rate to each STA is 2 spatial stream, MCS 7, 0.8-$\mu$s GI, 2$\times$ LTF. The BAs are retrieved using UL MU-MIMO with a per STA data rate allocation of 1 spatial stream, MCS 3, 3.2-$\mu$s GI, 4$\times$ LTF, and 20-MHz BW, which results in a data rate of 29.3 Mbps. A HE TB sounding sequence precedes every MU-MIMO sequence to retrieve CBF reports.

$$\text{Total airtime spent on backoff} = 4 \cdot 110.5\,\mu\text{s} \tag{3.58}$$

$$= 442\,\mu\text{s} \tag{3.59}$$

$$\text{Total airtime spent on data Tx} = 4 \cdot \left(36 + 12 + 4 \cdot 6.4 + \frac{80000 \cdot 8}{172}\right)\,\mu\text{s} \tag{3.60}$$

$$\approx 4 \cdot 3800\,\mu\text{s} \tag{3.61}$$

$$= 15200\,\mu\text{s} \tag{3.62}$$

$$\text{Duration of TB PPDU containing BAs} = 40 + 2 \cdot 12.8 + \frac{32 \cdot 8}{29.3}\,\mu\text{s} \tag{3.63}$$

$$\approx 81.6\,\mu\text{s} \tag{3.64}$$

$$\text{Duration of MU BAR trigger frame} = 36\,\mu\text{s} \tag{3.65}$$

$$\text{Total airtime spent on getting ACKs} = 4 \cdot (16 + 36 + 16 + 81.6)\,\mu\text{s} \tag{3.66}$$

$$= 598.4\,\mu\text{s} \tag{3.67}$$

The CBF report size for $\text{N}_\text{r} = 4$, $\text{N}_\text{c} = 2$, $\text{N}_\text{g} = 4$, 20 MHz BW and 6-bit precision is 3856 bits.

$$\text{Duration of TB PPDU containing CBF reports} = 40 + 2 \cdot 12.8 + \frac{3856}{29.3}\,\mu\text{s} \tag{3.68}$$

$$= 209.6\,\mu\text{s} \tag{3.69}$$

The total airtime consumed by one HE TB sounding sequence to two STAs is $(110.5 + 32 + 16 + 61.6 + 16 + 36 + 16 + 209.6) = 497.7\,\mu$s.

$$\text{Sounding airtime for all STAs} = 4 \cdot 497.7\,\mu\text{s} \tag{3.70}$$
$$= 1990.8\,\mu\text{s} \tag{3.71}$$
$$\text{Total airtime in MU-MIMO mode} = 1990.8 + 442 + 15200 + 598.4\,\mu\text{s} \tag{3.72}$$
$$= 18231.2\,\mu\text{s} \tag{3.73}$$

MU-OFDMA mode:

If the same traffic is sent using MU-OFDMA, each MU PPDU can serve eight STAs using an RU allocation of one 26-tone RU per STA. The PHY rate for 2 spatial stream, MCS 9, 26-tone RU with 0.8-$\mu$s GI is 23.6 Mbps. Assuming a maximum PPDU duration limit of 4,000 $\mu$s, this limits the maximum amount of traffic per STA in a single MU PPDU to approximately 11,500 bytes. So 80,000 bytes of DL traffic would require seven DL MU-OFDMA sequences.

$$\text{Total airtime spent on backoff} = 7 \cdot 110.5\,\mu\text{s} \tag{3.74}$$
$$= 773.5\,\mu\text{s} \tag{3.75}$$

The first six MU PPDUs will carry the maximum 11,500 bytes while the last MU PPDU will carry the remaining 11,000 bytes.

$$\text{Total airtime spent on data Tx} = 6 \cdot 4000 + \left(36 + 40 + 2 \cdot 6.4 + \frac{11000 \cdot 8}{23.6}\right)\mu\text{s} \tag{3.76}$$
$$= 27828.8\,\mu\text{s} \tag{3.77}$$

Let us assume the BAs are retrieved using UL MU-OFDMA with a per STA data rate of 3 Mbps, that corresponds to an allocation of 1 spatial stream, MCS 3, 26-tone RU with 3.2-$\mu$s GI and 4$\times$ LTF.

$$\text{Duration of MU BAR trigger frame} = 20 + \frac{(28 + 9 \cdot 8) \cdot 8}{24}\,\mu\text{s} \tag{3.78}$$

$$\approx 56\,\mu\text{s} \tag{3.79}$$

$$\text{Duration of TB PPDU containing BAs} = 40 + 12.8 + \frac{32 \cdot 8}{3}\,\mu\text{s} \tag{3.80}$$

$$= 148.8\,\mu\text{s} \tag{3.81}$$

$$\text{Total airtime spent on getting ACKs} = 7 \cdot (16 + 56 + 16 + 148.8)\;\mu\text{s} \tag{3.82}$$

$$= 1657.6\,\mu\text{s} \tag{3.83}$$

$$\text{Total airtime in MU-OFDMA mode} = 773.5 + 27828.8 + 1657.6\,\mu\text{s} \tag{3.84}$$

$$= 30259.9\,\mu\text{s} \tag{3.85}$$

The summary from (3.57), (3.73), and (3.85) is that it requires $24,060\,\mu$s in SU mode, $18,232\,\mu$s in MU-MIMO mode, and $30,260\,\mu$s in MU-OFDMA mode to send 80,000 bytes of DL traffic to the eight STAs. MU-MIMO is the best option for this example as it consumes the least airtime of $18,232\,\mu$s which is 24% lower than SU mode. MU-OFDMA is not appropriate for this example as it requires seven MU PPDUs to send all the traffic, thereby offsetting any savings in backoff time and ACKs. For this specific example, MU-OFDMA is 26% worse in airtime consumption compared to SU mode.

## 3.12 REFERENCES


[1] "IEEE Standard for Information Technology–Telecommunications and Information Exchange between Systems Local and Metropolitan Area Networks–Specific Requirements Part 11: Wireless LAN Medium Access Control (MAC) and Physical Layer (PHY) Specifications Amendment 1: Enhancements for High-Efficiency WLAN," *IEEE Std 802.11ax-2021 (Amendment to IEEE Std 802.11-2020)* (2021), pp. 1–767, DOI: 10.1109/IEEESTD.2021.9442429.

[2] "IEEE Standard for Information Technology–Telecommunications and Information Exchange between Systems - Local and Metropolitan Area Networks–Specific Requirements - Part 11: Wireless LAN Medium Access Control (MAC) and Physical Layer (PHY) Specifications," *IEEE Std 802.11-2020 (Revision of IEEE Std 802.11-2016)* (2021), pp. 1–4379, DOI: 10.1109/IEEESTD.2021.9363693.

[3] Evgeny Khorov et al., “A Tutorial on IEEE 802.11ax High Efficiency WLANs,” *IEEE Communications Surveys and Tutorials,* vol. 21. no. 1 (2019), pp. 197–216.

[4] Maddalena Nurchis and Boris Bellalta, “Target Wake Time: Scheduled Access in IEEE 802.11ax WLANs,” *IEEE Wireless Communications,* vol. 26. no. 2 (Apr. 2019), pp. 142–150.

[5] Francesc Wilhelmi et al., “Spatial Reuse in IEEE 802.11ax WLANs,” *arXiv:1907.04141* (Nov. 2019).

[6] Qiao Qu et al., “Survey and Performance Evaluation of the Upcoming Next Generation WLANs Standard - IEEE 802.11ax,” *Mobile Networks and Applications,* vol. 24 (Oct. 2019), pp. 1461–1474.

# Chapter 4

## Wi-Fi Security

Security is a key component of both wired and wireless networks. It is even more critical in the case of wireless networks as accessing them, unlike in wired networks, does not require physical access to the network jack or cable due to the broadcast nature of the wireless signals. It is difficult to confine the data-carrying radio waves to be within the physical security perimeter of the organization. To prevent unauthorized eavesdropping and/or tampering, data must be protected with confidentiality and integrity mechanisms. Furthermore, both users and providers must be assured of the correct identity of the communicating parties. Encryption and authentication are the two main security mechanisms employed to achieve these objectives.

Early Wi-Fi devices used a security protocol called wired equivalent privacy (WEP). WEP was designed to be at least as secure as a wired LAN and did not have the intention to achieve strong security. Very soon, researchers and hackers found multiple ways to break into WEP protocol and compromise confidentiality. Some of these attacks were made available to the public as tools to raise awareness of the vulnerabilities. WEP was a complete failure in terms of providing security [1], but this failure led to the development of today's secure Wi-Fi protocols. IEEE responded to these attacks by introducing the 802.11i amendment to security. However, this amendment took a long time to develop and be ratified. Since there was an immediate need to address the known vulnerabilities of WEP with the constraints of the existing hardware, the Wi-Fi Alliance (WFA) created an interim certification program called Wi-Fi protected access (WPA) in 2003 based on a subset of the 802.11i amendments. Later in 2004, WFA launched WPA2 certification that was fully compliant with 802.11i. WPA2 certification has been mandatory since 2006 for all Wi-Fi devices to bear the Wi-Fi trademark. To future-proof against threats from high-performance computers and to protect against offline dictionary attack, WFA introduced WPA3 certification in June 2018 and since July 2020 has made it a prerequisite for a device to get Wi-Fi certification. Although WEP and WPA have now been deprecated, these are described in the subsequent sections in addition to WPA2 and WPA3, so that the reader gets a historical perspective of the evolution of Wi-Fi security.

## 4.1 WIRED EQUIVALENT PRIVACY

WEP is the security protocol specified in the original IEEE 802.11 standard ratified in 1997. It leverages a preshared key (PSK) – a secret string of hexadecimal digits shared between the wireless access point (AP) and authorized users – known as the WEP key to provide both authentication and encryption. Access control is achieved by preventing stations without the correct WEP key from connecting to the AP. The data stream is encrypted leveraging the same WEP key and cannot be decrypted by anyone without the correct WEP key.

### 4.1.1 WEP Authentication

Authentication relies on a simple challenge-response protocol. The station (STA) starts by transmitting an authenticate request to the AP. In response, AP sends an authenticate challenge packet containing a random plaintext (unencrypted stream of bits) to the station. STA uses the preshared WEP key to encrypt the received plaintext and sends the result as payload in the authenticate response packet to the AP. Upon receiving this packet, AP checks if the STA used the correct WEP key to encrypt the plaintext. If the check passes, AP sends authenticate success to the STA, else it sends authenticate failure.

After successful authentication, STA associates with the AP. Once associated, the payload content of all subsequent data frames exchanged between AP and station are encrypted.

### 4.1.2 WEP Encryption

Encryption is done by exclusive-ORing the data stream with a pseudorandom stream of bits generated based on the WEP key using the RC4 algorithm. Two different key sizes are supported: 40-bit key represented using 10 hexadecimal digits (WEP-40) and 104-bit key represented using 26 hexadecimal digits (WEP-104). WEP uses RC4 encryption algorithm [2] because of its simplicity and low computational complexity. Furthermore, being a stream encryption algorithm that encrypts data as it flows, there is no need to hold the data in memory. RC4 is a symmetric key encryption. That is, both encryption and decryption use the same preshared secret key.

WEP uses a 32-bit cyclic redundancy check (CRC) for integrity check. It is computed based on all the bits in the unencrypted message and is appended to the end of the frame. Even if a single bit in the message gets corrupted or tampered with, the receiver will detect a CRC mismatch and reject the message. While this is very similar to conventional CRC, integrity check bits are computed and appended before encryption, whereas the conventional CRC is computed and appended after encryption.

### 4.1.3 WEP Vulnerabilities

Even though the WEP protocol allows assigning a unique key, known as the key-mapping key, for each AP-STA pair, system administrators often assign the same key, known as the default key, for all stations due to its ease. While using the default key, a STA can decrypt all other STA's messages. Moreover, the default key needs to be updated every time a member leaves the group to guarantee full security. For example, if someone leaves the company, in order to deny access to that member, the key on every other employee's device needs to be changed.

As Wi-Fi garnered more and more adoption, WEP came under the scrutiny of security experts. This led to the uncovering of several weaknesses listed below:

- Authentication is only one way and not mutual. In particular, the AP is not authenticated by the STA. Hence STA may associate to a rogue AP that spoofs the medium access control (MAC) address and SSID of an authentic AP.
- Since the same key is used for both authentication and encryption, any weakness in either can be exploited by an attacker.
- Both plaintext and its encrypted output are available during the authentication phase to an attacker allowing attackers to reverse engineer the PSK.
- The encryption key does not change over time.
- Management and control packets are not encrypted.

Attackers can exploit the above vulnerabilities to decrypt messages. Since the key does not change over time, WEP is prone to replay attack [2]. Furthermore, rogue devices can transmit management and control packets, which don't require any encryption, to clients resulting in a denial of service (DoS) attack [2]. These led to the development of new protocols described in the following sections.

## 4.2 WI-FI PROTECTED ACCESS

WPA was introduced in 2003 based on parts of the IEEE 802.11i amendment as an interim solution to address some of the vulnerabilities of WEP. It could be run on legacy WEP hardware with minor upgrades and was intentionally designed to be backward-compatible with WEP. This led to its rapid, hassle-free adoption.

WPA introduced a new default encryption method called temporal key integrity protocol (TKIP), while continuing to support WEP-40 and WEP-104 for backward compatibility. Two new authentication methods were also introduced. Because of the multiple options for encryption and authentication, a new information element (IE) called WPA IE was introduced for the AP and STA to exchange encryption and authentication capabilities supported. The WPA IE is present in beacon, probe request, probe response, association request, and association response packets.

In the following sections, two methods of authentication are described: WPA-Personal (also known as WPA-PSK) and WPA-Enterprise (also known as WPA-EAP). WPA-Personal protects unauthorized network access by utilizing a set-up password. WPA-Enterprise verifies network users through an authentication server.

### 4.2.1 WPA Personal

WPA-Personal relies on a 256-bit PSK shared between AP and all authorized STAs. In practice, STAs use a common password (passphrase) to get on the wireless network and PSK is derived as a function of the Wi-Fi password, SSID, and SSID length using a password-based key derivation function (PBKDF). Since everyone uses the same password, it is suitable when the network has few trusted users, like in homes and small businesses.

In this method, STA associates to the AP using 802.11 open system authentication. During this step, there is no real authentication per se; there is only an exchange of identity and capability. However, AP can leverage this exchange to enforce MAC address filtering based on an optional whitelist or blacklist of addresses specified by the network administrator. If a whitelist is specified, AP sends authorization response back only if the MAC address of the STA sending authorization request is on the list. Similarly, if a blacklist is specified, AP will not send authorization response back if the MAC address of the STA sending authorization request is on that list.

After successful association, a 256-bit shared secret key called pairwise master key (PMK) is generated by both AP and STA. In WPA PSK authentication, the PMK is the PSK. In addition to PMK, the AP generates a random 256-bit group master key (GMK).

AP and STA perform mutual verification of the PMK using a process called four-way handshake, as illustrated in Figure 4.1. In addition to two-way (or mutual) verification of the PMK credentials, this handshake generates the pairwise transient key (PTK) used to encrypt unicast traffic and the group temporal key (GTK) used to encrypt broadcast and multicast traffic.

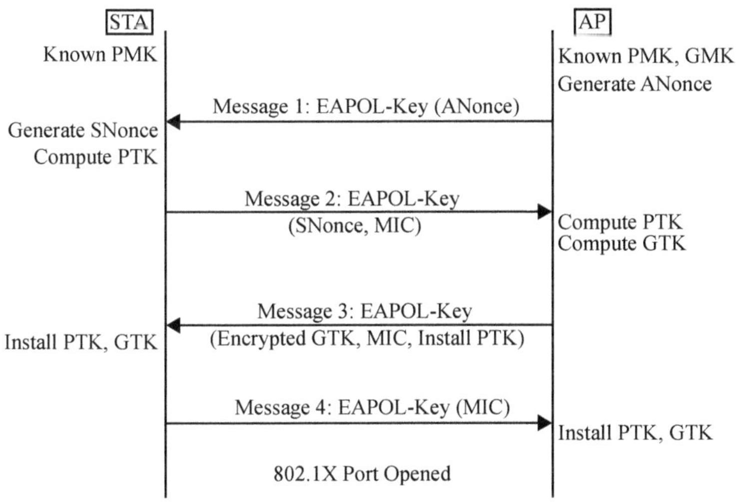


**Figure 4.1** Four-way handshake.

When the four-way handshake starts, both AP and STA have the knowledge of PMK and AP additionally has the knowledge of GMK. Both AP and STA independently generate a random number called ANonce and SNonce, respectively (nonce in cryptographic lingo is a random number used once).

AP starts by sending its ANonce to STA. Upon receiving, STA computes PTK as a function of PMK, ANonce, SNonce, and the MAC addresses of AP and STA. Then, STA sends its SNonce to AP. Based on this, AP computes PTK as the same function of PMK, ANonce, SNonce, and the MAC addresses of AP and STA. In addition to this, AP computes GTK as a function of GMK without any involvement of the STA. As a next step, AP encrypts GTK using PTK and sends it to STA. Upon receiving, STA decrypts the message using its copy of PTK, retrieves GTK, and sends an ACK back to the AP. At the end of this four-way handshake, both AP and STA have mutually verified PMK and have obtained PTK and GTK. Until the four-way handshake is completed successfully, only EAP over LAN (EAPOL) frames are allowed and the client is not granted network connectivity despite being associated.

Since all broadcast and multicast traffic are encrypted using the same GTK, an attacker could potentially guess the key by sniffing enough broadcast and multicast packets. To prevent this, GTK is updated, or rotated, periodically and also when a device leaves the network. The 802.11i amendment [3] specifies a two-way handshake protocol to handle the GTK rotation.

### 4.2.2 WPA Enterprise

WPA Enterprise is based on the 802.1X protocol. A remote authentication dial-in user service (RADIUS) authentication server, which runs either on the AP itself or on a dedicated machine when there are multiple APs in the network, is used to authorize stations for network access. WPA Enterprise is appropriate when there are a large number of users, making it practically impossible to safeguard the common password from unauthorized users. Network administrators can choose whether to use digital certificates or credentials to authenticate users. Certificates offer a superior form of security compared to credentials as the user never has to enter a password to reconnect and they cannot be stolen by an outside attacker.

The process starts with STA ("supplicant"), AP ("authenticator"), and authentication server exchanging authentication messages using extensible authentication protocol (EAP). Enterprises have the flexibility to choose from a range of authentication protocols depending on their requirements for security level, authentication speed, and ease of maintenance. Some commonly used authentication protocols are: EAP-Transport Layer Security (EAP-TLS), EAP-Tunneled Transport Layer Security (EAP-TTLS), and Protected Extensible Authentication Protocol (PEAP). The EAP packets from STA to AP are encapsulated using the EAP over LAN (EAPOL) protocol, while the EAP packets from AP to authentication server are encapsulated using the RADIUS protocol as shown in Figure 4.2.

Successful authentication results in generation of a unique PMK between STA and authentication server for each session leveraging the famous Diffie Hellman exchange [4]. The PMK is sent by authentication server to AP in a secure way

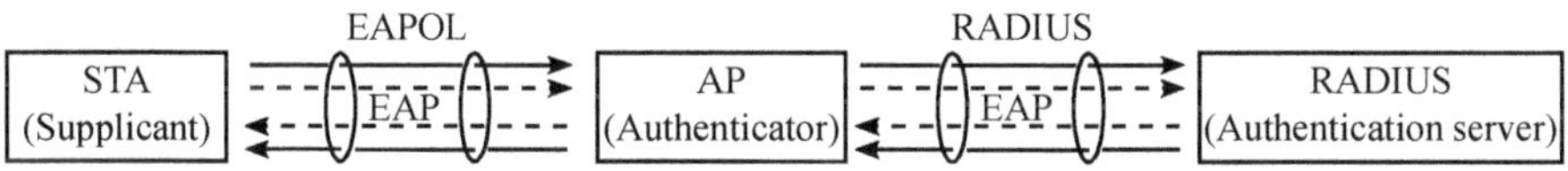


**Figure 4.2** EAP authentication process.

leveraging a shared key between AP and the authentication server. AP also generates a random 256-bit GMK. Then, the AP and STA follow the four-way handshake protocol described in Section 4.2.1, to establish PTK and GTK that are used to encrypt traffic. WPA-Enterprise performs GTK rotation periodically as well as when a device leaves the network using two-way handshake protocol.

Some consumer electronic devices, such as game consoles and video streaming devices, do not support Enterprise (802.1x or EAP) mode of authentication necessitating the use of WPA2-Personal option.

### 4.2.3 WPA Encryption

WPA encryption (TKIP) leverages the same RC4 encryption algorithm defined for WEP, but uses a 128-bit key to eliminate the weakness associated with short WEP keys. Another important feature of TKIP is that it uses a different encryption key for every data packet, making it difficult for hackers to decrypt messages. This is achieved by combining the 48-bit sequence number of the packet, along with PTK and the MAC address of the transmitter, to dynamically generate the key used to initialize the RC4 algorithm. The introduction of sequence number, which gets incremented every time a new packet is transmitted, also helps to protect against replay attack. For all practical purposes, the 48-bit sequence number does not repeat itself during a communication session and hence any attempt to replay old packets will be detected as out-of-order and dropped by the receiver.

### 4.2.4 Message Integrity Check

WPA replaces the 32-bit CRC of WEP with a stronger 64-bit message integrity check (MIC) called Michael, which could be implemented using a relatively low power processor without requiring any high-speed hardware multiply. The computation of Michael MIC requires only substitutions, rotations, and exclusive-OR operations on the MSDU, which consists of the actual payload, and source and destination addresses. The 64-bit Michael is appended to the MSDU prior to fragmentation and encryption.

## 4.3 WPA2

Introduced in 2004, WPA2 is the Wi-Fi Alliance's name for the 802.11i security amendment. While WPA incorporates only a subset of 802.11i, WPA2 is in full compliance with 802.11i. WPA2 significantly improves security by mandating advanced encryption standard (AES) encryption, which is National Institute of Standards and Technology (NIST) FIPS 140-2 compliant. WPA2 replaces Michael MIC with cipher block chaining message authentication code (CBC-MAC) MIC.

WPA2 includes several optional features: PMK caching and preauthentication to make reassociation and roaming faster, Wi-Fi protected setup (WPS) to simplify authentication for home users, and protected management frame (PMF) to minimize vulnerability to DoS attack.

WPA2 uses robust secure network (RSN) IE to exchange encryption and authentication capabilities between AP and STA. The RSN IE is present in beacons, probe request, probe response, association request, and association response frames.

### 4.3.1 WPA2 Authentication

Like WPA, WPA2 supports two version of authentication: WPA2-Personal, which relies on a common password for authentication, and WPA2-Enterprise, which uses an authentication server for authentication. WPA2-Personal and WPA2-Enterprise use the same authentication steps as in WPA-Personal and WPA-Enterprise, respectively. At the end of the four-way handshake, both AP and STA have PTK and GTK used to encrypt and decrypt traffic. Refer to the website [5] for some examples of wireless sniffer traces that help show the message exchanges in detail.

### 4.3.2 WPA2 Encryption

While WEP and WPA use RC4 algorithm for encryption, WPA2 uses a more advanced and complex algorithm called AES shown in Figure 4.3. The specific protocol used is called AES-CCMP, where CCMP stands for Counter mode CBC-MAC Protocol. Like RC4, AES is a symmetric key algorithm that uses the same key for both encryption and decryption. Unlike RC4, AES encrypts data in discrete 128-bit sized blocks rather than encrypting data as it streams. Because of the block size used, it is often referred to as AES-CCMP-128 encryption. The encryption algorithm groups each block of 128 bits into 16 bytes, arranges the 16 bytes as a $4 \times 4$ matrix, and performs a four-step process. The four steps are substitution of bytes, circular shift of rows, transformation of columns using a mathematical function, exclusive-OR with keys derived from PTK or GTK, along with packet number, using a process called key expansion. These four steps are iterated 10 times to arrive at the encrypted result.

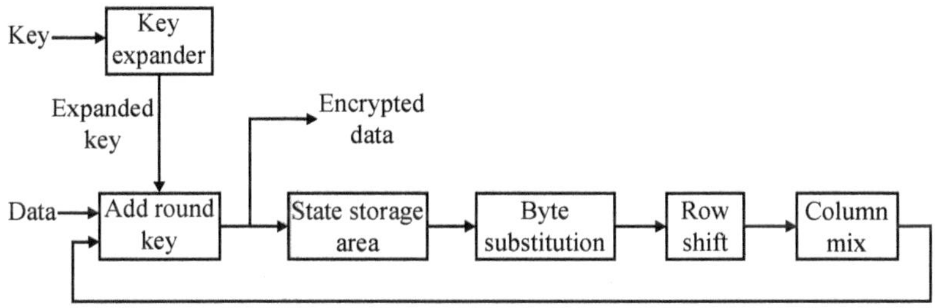


**Figure 4.3** AES encryption.

While the RC4 decryption function is identical to the encryption function, the AES decryption function is different than the encryption function since AES decryption requires repeating the encryption steps but in reverse order. Because of the complexity involved, AES encryption and decryption are implemented using a custom hardware in order to meet the data rate requirements.

### 4.3.3 Message Integrity Check

Like WPA, AES uses a MIC algorithm to ensure that the packet has not been tampered with. However, it is different than the Michael algorithm used in WPA. AES MIC based on CBC-MAC algorithm operates on 128-bit blocks like AES encryption and the result is carried forward from one block to the next until the end of the message is reached. Then the final 64-bit value is calculated and appended to the encrypted message.

### 4.3.4 PMK Caching

While using WPA2-Enterprise, when a STA associates to an AP, it carries out the EAP exchange followed by a four-way handshake. When the PMK caching feature of WPA2 is enabled, the AP caches the PMK identifier from the EAP exchange. Upon subsequent reassociations by the same client, the EAP exchange is skipped and the client directly proceeds to four-way handshake leveraging the cached PMK. If a client connected to an AP either roams away or loses association and returns again to reassociate, PMK caching decreases authentication time. There is a configurable session timeout value in RADIUS to set a session expiration time after which the cached keys would be cleared.

### 4.3.5 Opportunistic Key Caching

Opportunistic key caching (OKC) is an extension of PMK caching. It is a non-standard feature used in controller-managed networks with several APs. Here, all APs managed by the same controller will receive a copy of the PMK identifier of every client in the network. Some implementations may store this information in the controller itself. This enables client devices already authenticated via 802.1X to directly proceed to a four-way handshake while roaming to other APs in the same network.

### 4.3.6 Fast BSS Transition

Fast BSS transition (FT) is a standards-based alternative, introduced as part of the 802.11r amendment, to OKC for fast, secure roaming. In addition to avoiding the 802.1x EAP exchange like OKC does, FT roaming as illustrated in Figure 4.4 avoids the regular four-way handshake as well by piggybacking onto the newly introduced FT authentication and FT reassociation messages that replace the standard authentication and association messages. This makes FT roaming even more efficient than OKC. One of the issues with 802.11r is that many older client devices don't support it, and even have trouble properly detecting and associating to networks with 802.11r enabled.

### 4.3.7 Protected Management Frames

Neither management nor control frames are encrypted while using WEP or WPA. This makes the network vulnerable to DoS attack. The PMF feature, included as part of WPA2, is based on the 802.11w amendment and provides message integrity protection for unicast and broadcast management frames, and also encrypts unicast management frames to provide confidentiality. PMF protects against disassociation and deauthentication type of DoS attacks. It applies the same encryption done on unicast data frames to unicast management frames. Examples of unicast management frames encrypted are: disassociation, deauthentication, blockack, radio measurement action frame (802.11k frames), QoS action frame, channel switch announcement (CSA), and BSS transition management (BTM) frames. PMF uses the broadcast integrity protocol (BIP) to provide data integrity and replay protection

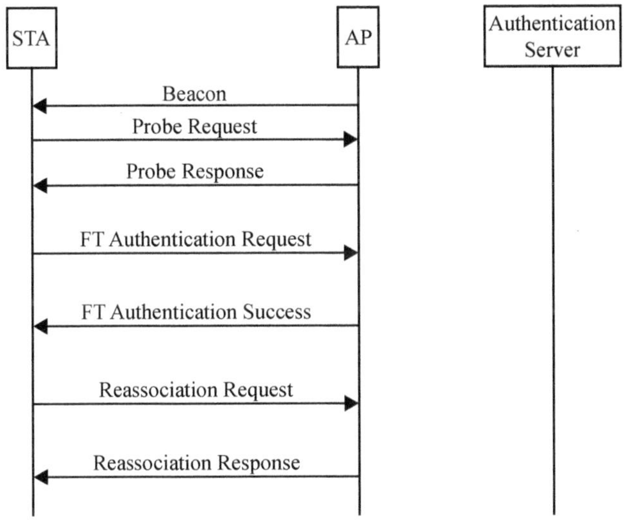


**Figure 4.4** Fast BSS transition roaming.

to broadcast management frames by appending a MIC to all broadcast management frames. One limitation of PMF is it does not provide any protection to frames sent before the successful completion of a four-way handshake.

In practice, many Wi-Fi networks do not enable PMF by default since some legacy STAs (Wi-Fi 4 and earlier) do not support this feature. However, APs typically allow PMF to be configured as optional, in which case the AP will apply PMF only to STAs that advertise support for it. Support for PMF is mandatory since Wi-Fi 5 certification.

### 4.3.8 Wi-Fi Protected Setup

Wi-Fi protected setup (WPS) is a feature introduced in WPA2 to make device provisioning faster and easier. It is a useful method for novice home users to set up a Wi-Fi connection quickly without typing in, or even knowing, the SSID or password. This is done in one of two ways: PIN method, where the user chooses a unique 8-digit PIN for each STA and enters the same on the AP during authentication, and push button method, where the user pushes a physical button on both and STA within a 2-minute time window. Beacons, probe requests, and probe responses of WPS-capable devices have an additional Wi-Fi simple configuration (WSC) IE to indicate WPS support.

The push button method is more popular compared to the PIN method. Since 2011, the PIN method has not been recommended due to the discovery of a vulnerability. Though the push button has a chance to allow unauthorized STAs to connect since any STA that attempts to join AP during the time window after the button push will be accepted, it is still preferred by many home users for its simplicity. However, one practical hurdle is that not all devices support WPS as it is an optional feature.

## 4.4 KEY REINSTALLATION ATTACK

In 2017, more than a decade after WPA2 came into existence, researchers [6] uncovered security vulnerabilities in the four-way handshake used by both WPA2-Personal and WPA2-Enterprise authentication protocols.

WPA2 is designed to encrypt different packets using different keys by including the packet number in key generation, but an attacker can exploit a vulnerability in the four-way handshake implementation to trick the AP into using the same key to encrypt multiple packets. In case one of these packets has a known content, other packets encoded using the same key can be easily decrypted. However, it is harder to decrypt packets when there is no known content, although it is easy to find packets with known content in practice. If the traffic is encrypted using higher layer protocols such as https, VPN and TLS, the attacker cannot decrypt the data even if they succeed in decrypting the Wi-Fi packet.

The key reinstallation attack (KRACK) vulnerability does not point to any fundamental issue with the security properties of the WPA2 protocol. The confidentiality of neither the encryption key nor the password is ever compromised. The vulnerability exploits the protocol's failure to explicitly prohibit retransmission of packets during the four-way handshake. So, the KRACK vulnerability can be addressed using a software patch, which most device vendors released shortly after this vulnerability was made public. However, it caused a fresh wave of security concerns among Wi-Fi users leading to the introduction of WPA3.

## 4.5 WPA3

To future-proof Wi-Fi against security threats from high performance computers and to win back user confidence in the aftermath of KRACK, WFA announced a new certification program called WPA3 [7] in 2018. All Wi-Fi 6 devices support WPA3. Below are the key enhancements introduced in WPA3, which continues to support the AES encryption specified in WPA2:

- Personal (PSK) mode of authentication is replaced by simultaneous authentication of equals (SAE) to protect against dictionary attack.
- Enterprise (EAP) mode of authentication is strengthened by the addition of an optional 192-bit encryption.

Furthermore, WPA3 mandates support for both PMK caching and PMF, as well as the software fix for the KRACK vulnerability. Since WPA3 supports the encryption algorithms in WPA2, most WPA2 capable hardware can be software upgraded to support WPA3.

### 4.5.1 WPA3 Authentication

WPA3 supports two versions of authentication: WPA3-SAE, which replaces WPA2-Personal while continuing to rely on a common password, and WPA3-Enterprise, which uses a network server for authentication.

Even though WPA and WPA2-Personal use a 256-bit PSK, as mentioned in Section 4.2.1, the PSK is usually derived from a common password, which is often short, making it vulnerable to an offline dictionary attack. Here, the attacker captures AP SSID, MAC address of AP and STA, and ANonce and SNonce from the first two messages of the four-way handshake, and uses an offline guess-and-check technique to crack the password. The attacker starts by guessing a password and computes the corresponding PMK and PTK leveraging captured information such as SSID, MAC address, and Nonce. The second message of the four-way handshake includes a MIC, which is a function of PTK. This allows the attacker to check if the MIC computed using the PTK based on the guessed password matches the MIC in the captured message. A match corroborates that the guessed password is indeed correct. If there isn't a match, attacker guesses a new password and repeats the above steps until there is a match.

The reason for the above vulnerability in WPA2-PSK is that the password search space is significantly smaller than the PMK search space and the four-way handshake contains information required to check whether the guessed password is correct or not. This means the whole password guess-and-check can be done completely offline in a relatively short time, especially using a high-performance computer. Once the password is compromised, there is no confidentiality for any user of the network. To mitigate this risk, WPA3 replaces PSK with SAE.

SAE is resistant to dictionary attacks by forcing the attacker to interact with either the AP or STA to make guesses rather than being able to identify the use of a password in their dictionary using an offline or passive computation. SAE authentication maps the password to a group element and then uses the elliptic curve Diffie Hellman (ECDH) message exchange to generate a unique PMK between AP and STA for every session. Following PMK generation, the four-way handshake is completed before allowing network connectivity. In simple terms, SAE authentication is achieved by devices proving to each other that they know the correct password without revealing the password or any reverse-engineer-able mapping of it. This proving process requires "online" interactions with the devices, making it time consuming and rendering any attempt to attack easy to detect. Furthermore, in the unlikely event that the password gets compromised, the attacker cannot decrypt any other STA's data because the SAE protocol generates a unique PMK (also known a personalized key) for every session even though a common password is used by everyone. Figure 4.5 illustrates the message exchanges involved in SAE authentication.

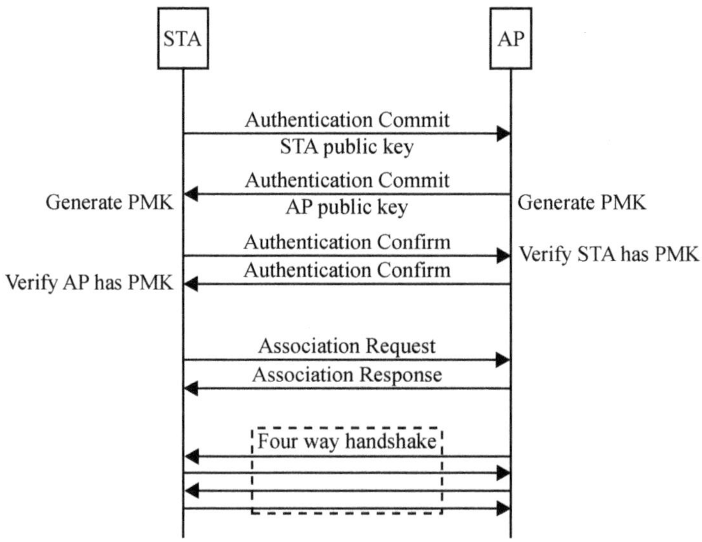


**Figure 4.5** WPA3-SAE authentication.

WPA2-Enterprise is much more secure compared to WPA2-Personal. Hence, WPA3-Enterprise does not bring in any significant change other than an optional support for longer 192-bit key security. WPA3-Enterprise mandates using server certificate validation when a RADIUS server is used for authentication so that the STA can validate the server certificate and confirm that it is connecting to the correct RADIUS server. This prevents the STA from sending its credentials to a rogue AP.

### 4.5.2 WPA3 Encryption

In addition to the AES-CCMP-128, WPA3 optionally adds support for a stronger 256-bit key AES Galois Counter Mode Protocol (GCMP) encryption. AES-GCMP-256 is specified as an optional feature in 802.11ac to meet its higher data rate requirements.

GCMP has computational advantages over CCMP as it requires only half the number of encryption operations for the same security level. CCMP is based on a chained mode of operation that requires the output of one stage to be used as input to the next, making it difficult to parallelize the computations. On the other hand, due to lack of chaining, GCMP is amenable to parallelization, making it a better candidate for Wi-Fi 5 and Wi-Fi 6 owing to their higher data rates.

### 4.5.3 WPA3 Transition Mode

WFA also supports a WPA3 transition mode or mixed mode that allows both WPA3 and WPA2 to be configured on the same BSS. This mode enables both WPA3 and WPA2 capable clients to connect to the same BSS. WPA3 authentication method is applied to WPA3 capable clients, while WPA2 authentication method is used with WPA2 capable clients. Similarly, PMF is mandatorily enabled for WPA3 capable clients while it is optional for WPA2 capable clients.

### 4.5.4 Device Provisioning Protocol

Device provisioning protocol (DPP) is proposed as a more secure replacement for WPS. DPP is part of a separate WFA certification program called Wi-Fi easy connect although it was initially planned to be part of WPA3. It is especially useful for configuring Internet of Things (IoT) devices, which are often headless devices that don't have keyboard or display user interfaces. DPP uses another device, such as mobile phone, to get these headless devices connected securely employing various methods such as Near-Field Communication (NFC), password, and Quick Response (QR) code.

The devices to be configured are called enrollees. Enrollees could be clients or APs. Configuration is often done using an application running on a smartphone device playing the role of configurator. A configurator establishes a Wi-Fi direct link with the enrollee and the configuration is done wirelessly and securely using the DPP protocol. DPP defines three phases: authentication, configuration, and network access.

In the authentication phase, configurator reads a QR code printed physically on the enrollee containing a public key for encryption, MAC address of the enrollee, and the channel where the enrollee accepts communication. Based on this information from QR code, configurator first establishes an open encryption Wi-Fi direct link with the enrollee. This can also be an NFC or Bluetooth link instead of the Wi-Fi direct link. Configurator and enrollee exchange packets during this phase to generate the key to be used for securing the messages in configuration phase.

During the configuration phase, configurator uses the key generated in the authentication phase to securely transmit to the enrollee either the SAE or WPA2-PSK password or a certificate called DPP connector that can be used to detect other enrollees configured by the same configurator. The network access is straightforward when the enrollee has the SAE or WPA2 password, which it can use to connect with another enrollee configured with the same password. If the enrollee has a DPP connector instead, network access phase involves some more steps. The enrollee (client) first discovers another enrollee (AP) with the same DPP connector. Once both verify the digital signature to confirm that both were configured by the same configurator, they exchange packets to generate a shared secret, which in turn is used to generate PMK. Then the usual four-way handshake follows resulting in the client establishing connection with the AP.

### 4.5.5 Opportunistic Wireless Encryption

Opportunistic wireless encryption (OWE) is a new feature that provides unauthenticated encryption for open WLANs. Open wireless LANs, also known as open SSIDs, are widely used in locations where user authentication is deemed unnecessary or distribution of credentials is impractical. Examples of such locations include public places such as coffee shops and restaurants, and guest networks with a web portal in places like hotels and airports. Despite its convenience due to ease of connection, there is no protection of data over the air, allowing attackers to read the user data although there could be other higher-layer protections such as https, VPN, and TLS that prevent attackers from deciphering the data.

OWE provides protection against passive eavesdropping for open networks without requiring a user to do anything extra, thereby retaining the ease of use. OWE-capable devices exchange public key messages in association request and association response frames using the EDCH exchange to generate a unique PMK between AP and STA. Then, the usual four-way handshake follows to establish an encrypted link between the AP and STA.

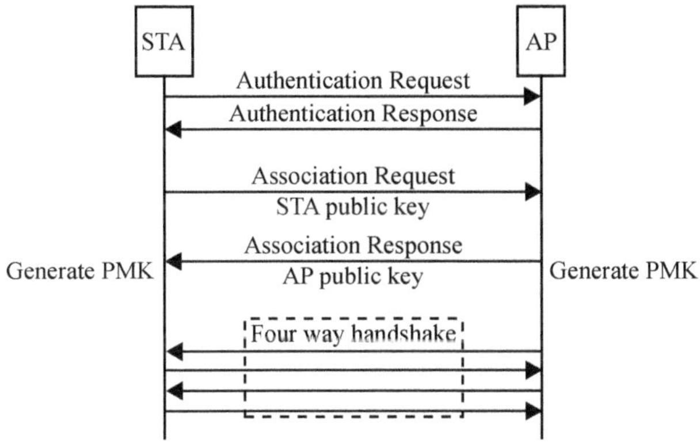


**Figure 4.6** OWE connection process.

To ensure a smooth transition for legacy clients that do not support OWE, an OWE-capable AP advertises two beacons:

- Beacon 1: This advertises open security and is meant for legacy clients. This beacon includes an OWE transition mode element (OTME), which points to the SSID and BSSID of beacon 2.
- Beacon 2: This is a hidden SSID meant for OWE-capable clients. The OTME in this beacon points to the SSID and BSSID of beacon 1. This beacon also includes an RSN IE that advertises the encryption algorithms supported.

OWE is part of a separate WFA certification program called Wi-Fi Enhanced Open although it was initially planned to be part of WPA3. Since the OWE protocol does not include any two-way authentication, although AP has the option to authenticate STAs through other higher layer protocols such as captive portal, it is prone to a rogue AP attack [8].

## 4.6 MULTICHANNEL MAN-IN-THE-MIDDLE

Several types of attacks require the ability to block, modify, or delay encrypted frames sent between a client and AP. In 2014, Vanhoef and Piessens proposed a multichannel man-in-the-middle (MitM) technique [9] that enables an adversary to perform such malicious acts easily using commodity hardware. In this multichannel MitM technique, an attacker AP capable of dual channel concurrent operation clones a legitimate AP on a different channel and forces the client to initiate connection to the attacker AP. However, the attacker AP takes a MitM position by simply forwarding frames between the client and legitimate AP on the two different channels. Essentially, the client is authenticated and connected to the legitimate AP but through the attacker AP. From the perspective of the legitimate AP, the attacker AP is like a client connected to it. Although the attacker AP cannot decrypt frames from the client by taking a MitM position, it can block, modify, or delay encrypted frames sent between client and legitimate AP. Figure 4.7 depicts the multichannel MitM technique.

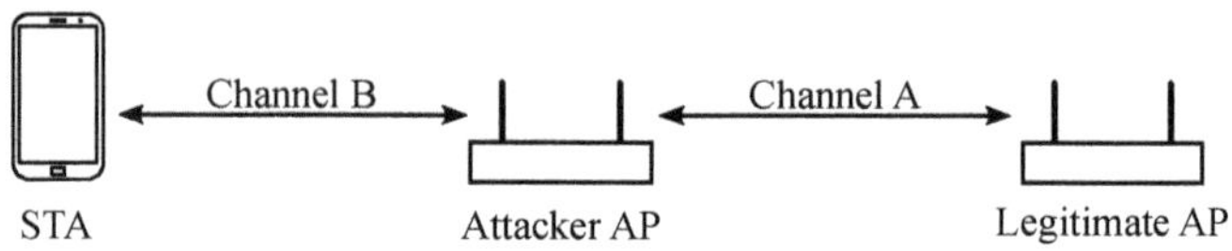


**Figure 4.7** Multichannel man-in-the-middle technique.

The multichannel MitM technique is a simple yet powerful technique that enables attackers to develop various types of attacks. The KRACK attack is a good example of how multichannel MitM can be used to exploit an implementation vulnerability in four-way handshake protocol. Recognizing the threat of MitM based attacks, the IEEE standard [3] has added two new features called operating channel validation (OCV) and beacon protection to prevent MitM attack based on the solution described in [10, 11]. Since December 2020, WFA has also updated the WPA3 certification program [12] to include OCV and beacon protection as optional features.

### 4.6.1 Operation Channel Validation

OCV adds a new operating channel information (OCI) IE containing the primary channel number and bandwidth information in all security-related Wi-Fi exchanges such as four-way handshake, FT roaming, and channel switch announcement messages. This enables both client and AP to mutually authenticate the channel specified in OCI against the current operating channel. If the operating channel validation fails, the handshake messages are simply dropped, thereby preventing an attacker AP from taking a multichannel MitM position.

### 4.6.2 Beacon Protection

The beacon protection feature provides integrity protection to beacon frames by adding a MIC element that covers all contents except the timestamp field. The beacon integrity key is provisioned to clients during the four-way handshake. This feature enables associated clients to detect any manipulation of beacon frame contents by an attacker AP.

## 4.7 FRAGMENTATION AND AGGREGATION ATTACKS

In [13], Mathy Vanhoef presented several new Wi-Fi attacks using the multichannel MitM technique to compromise the confidentiality of sensitive data by targeting design flaws in the implementation of Wi-Fi defragmentation and AMSDU deaggregation functionalities. A vast majority of Wi-Fi client and AP devices including the latest WPA3 compliant devices were shown to be vulnerable to these new set of attacks broadly referred as fragmentation and aggregation attacks (FragAttacks). FragAttacks were made public on May 11, 2021 by the WFA along with several recommendations to address them. Most device vendors have released software updates to address the fragmentation and aggregation vulnerabilities.

The MAC header of any encrypted frame is always sent in plaintext but the MIC provides integrity protection to most contents of the MAC header and this enables detection of any forgery by a MitM. However, certain fields of the MAC header are not fully authenticated, implying any manipulation of these fields by a MitM cannot be detected. In particular, the sequence number field and the AMSDU Present subfield of QOS control field (see Figure 4.8) are not authenticated.

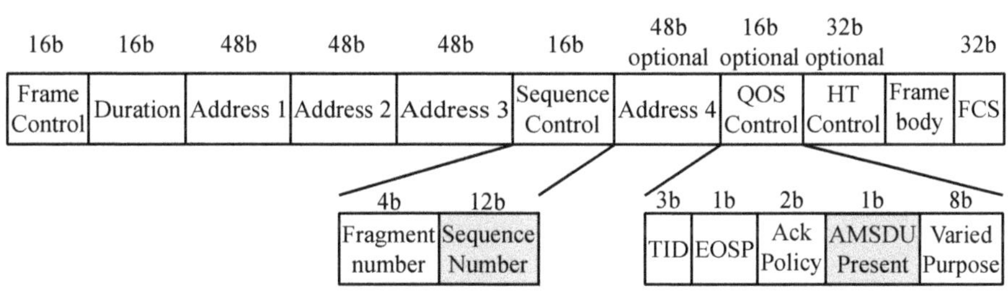


**Figure 4.8** Unauthenticated MAC header fields.

The fragmentation attack uses multichannel MitM technique to first block a certain fragmented encrypted frame and then later forge the sequence number field of another fragmented frame encrypted with a different key. Encryption keys can change due to various reasons such as a periodic rekey event or upon reassociation. Since the victim device (client or AP) is unaware of the sequence number manipulation by the MitM attacker, it causes the victim to decrypt these two fragments, reassemble them, and route it according to the contents of the resultant reassembled packet. In theory, all fragments of a frame must be encrypted by the same key, but most Wi-Fi implementations do not check for this during defragmentation. This weakness in defragmentation implementation is abused by the fragmentation attack to inject arbitrary packets into the network or route sensitive data to an attacker controlled server. The fragmentation attack can be prevented by a software change in defragmentation implementation to not reassemble fragments of two different security contexts and also to clear the fragment cache whenever the security context changes such as a rekey or reassociation event. An interesting fact about the fragmentation attack is that it does not break the Wi-Fi password or the encryption keys but still manages to make the victim decrypt and forward sensitive data to an attacker controlled destination. This is yet another example and reminder of the threat posed by multichannel MitM technique.

The aggregation attack involves an attacker AP taking a multichannel MitM position as shown in Figure 4.7 and forging the unauthenticated AMSDU present subfield of certain frames causing the victim device (AP or client) to incorrectly treat the frame's payload as AMSDU subframes. Although this causes the first AMSDU subframe to have an unknown sender and destination MAC address, most Wi-Fi implementations proceed to process the subsequent AMSDU subframes. This implementation weakness in AMSDU deaggregation is abused by the aggregation attack to inject arbitrary packets into the network. The ability to inject arbitrary packets enables an attacker to trick the victim into using an attacker controlled domain name system (DNS) server or perform a port scan on the victim. This attack can be mitigated by dropping all subframes in an AMSDU if the first subframe has an unknown sender or destination MAC address. An ideal solution to this problem would be to use the signaling and payload protection AMSDU feature of 802.11 standard [3] but this feature is not supported by most Wi-Fi devices.

## 4.8 ROGUE AP DETECTION

Any unauthorized AP deployed in a secure network is referred to as a rogue AP. Most enterprise-grade APs implement a few simple steps to detect and prevent rogue AP attack. This feature is often referred to as wireless intrusion detection system (WIDS) and wireless intrusion prevention system (WIPS) [14].

Rogue AP detection [14] works by having APs periodically scan all channels either passively or actively by sending broadcast probe requests. The AP checks if any of the probe responses or beacons contain either an unexpected SSID or an unexpected BSSID with legitimate SSID based on its knowledge of the SSIDs and BSSIDs of the authorized APs in the network. Any unexpected SSID and/or BSSID points to the presence of rogue APs. Some implementations may have some shared secret among the authorized APs in a network which can aid rogue AP detection.

In addition, a passive scan helps to check if there is any frame with a legitimate BSSID from a client that is not associated to any of the authorized APs in the network. The presence of such frames implies that the client is associated to a rogue AP. Then, the AP sends a deauthentication frame to such strayed-away clients masquerading as the rogue AP so that they disconnect from the rogue AP and a disassociation frame to rogue AP masquerading as strayed-away clients so that the rogue AP disconnects them. The disassociation frame to rogue APs prevents clients in power save state to resume connection with rogue AP as they may have missed the deauthentication frame from AP.

## 4.9 SUMMARY

**Table 4.1**
Evolution of Wi-Fi Security

| | *WEP* | *WPA* | *WPA2* | *WPA3* |
|---|---|---|---|---|
| Certification start | | 2003 | 2004 | 2018 |
| Mandatory since | | | 2006 | 2019 |
| Underlying standard(s) | 802.11 | 802.11i (Partial compliance) | 802.11i, 802.11w(PMF) | 802.11i, 802.11s(SAE), 802.11w(PMF) |
| Information element | None | WPA IE | RSN IE | RSN IE |
| Encryption | RC4 | TKIP | TKIP, AES-CCMP | AES-CCMP, AES-GCMP |
| Authentication | None | PSK, 802.1X | PSK, 802.1X | SAE, 802.1X |
| Usability enhancement | None | None | WPS | DPP |
| Open Wi-Fi security | None | None | None | OWE |

Table 4.1 summarizes the evolution of Wi-Fi security over the years. While the early implementations based on WEP had several vulnerabilities, they have been addressed with the introduction of WPA2 that remained secure for about 14 years. No significant security changes were mandated during those 14 years primarily because of the reluctance to touch something that is not broken. With the discovery of KRACK vulnerability, new security certification called WPA3 has been introduced. WPA3 includes new features that future-proof Wi-Fi against attacks using high-performance computers and also protect Wi-Fi against offline dictionary attack. Even open networks have been secured with the introduction of OWE.

## 4.10 REFERENCES

[1] W. A. Arbaugh et al., "Your 80211 wireless network has no clothes," *IEEE Wireless Communications,* vol. 9. no. 6 (Dec. 2002), pp. 44–51.

[2] J. Edney and W. Arbaugh, *Real 802.11 Security: WiFi Protected Access and 802.11i*, Addison-Wesley, 2004.

[3] "IEEE Standard for Information Technology–Telecommunications and Information Exchange between Systems - Local and Metropolitan Area Networks–Specific Requirements - Part 11: Wireless LAN Medium Access Control (MAC) and Physical Layer (PHY) Specifications," *IEEE Std 802.11-2020 (Revision of IEEE Std 802.11-2016)* (2021), pp. 1–4379, DOI: 10.1109/IEEESTD.2021.9363693.

[4] H. X. Mel and D. Baker, *Cryptography Decrypted*, Addison-Wesley, 2005.

[5] Rasika Nayanajith, 2021, URL: https://mrncciew.com/.

[6] Mathy Vanhoef and Frank Piessens, "Key Reinstallation Attacks: Forcing Nonce Reuse in WPA2," *Proceedings of the 24th ACM Conference on Computer and Communications Security*, Nov. 2017.

[7] Wi-Fi Alliance, *Wi-Fi Alliance WPA3 Specification*, 2020, URL: https://www.wi-fi.org/file/wpa3-specification.

[8] Aaron E. Earle, *Wireless Security Handbook*, Auerbach Publications, 2006.

[9] Mathy Vanhoef and Frank Piessens, "Advanced Wi-Fi attacks using commodity hardware," *Proceedings of the 30th Annual Computer Security Applications Conference*, Dec. 2014.

[10] Mathy Vanhoef et al., "Operating channel validation: Preventing multi-channel man-in-the-middle attacks against protected Wi-Fi networks," *Proceedings of the 11th ACM Conference on Security and Privacy in Wireless and Mobile Networks*, June 2018.

[11] Mathy Vanhoef, Prasant Adhikari, and Christina Pöpper, "Protecting Wi-Fi beacons from outsider forgeries," *Proceedings of the 13th ACM Conference on Security and Privacy in Wireless and Mobile Networks*, July 2020.

[12] Wi-Fi Alliance, *Wi-Fi Alliance Wi-Fi Security Roadmap and WPA3 Updates*, Dec. 2020, URL: https://www.wi-fi.org/downloads-public/202012_Wi-Fi_Security_Roadmap_and_WPA3_Updates.pdf.

[13] Mathy Vanhaoef, "Fragment and Forge: Breaking Wi-Fi Through Frame Aggregation and Fragmentation," *Proceedings of the 30th USENIX Security Symposium*, Aug. 2021.

[14] Cisco Meraki, *White Paper Air Marshal*, Sept. 2013, URL: https://meraki.cisco.com/lib/pdf/meraki_whitepaper_air_marshal.pdf.

# Chapter 5

# Wi-Fi Implementation

A Wi-Fi implementation has to meet several requirements depending on the specific use case while also complying to the IEEE 802.11 standard [1, 2], country-specific regulations, and Wi-Fi Alliance (WFA) certification test plan. The choice of hardware components and software algorithms in a Wi-Fi implementation vary a lot depending on the application. For example, a Wi-Fi based Internet of Things (IoT) device and a Wi-Fi access point serving hundreds of STAs have completely different set of objectives. This chapter intends to throw some light into the various design choices and challenges involved in implementation that can aid in better Wi-Fi system design and qualification. The objective of this chapter is to empower those involved in Wi-Fi implementation to ask the right questions and be aware of the common challenges.

## 5.1 HARDWARE

Hardware determines the boundaries of what is possible with software. Certain requirements of Wi-Fi cannot be addressed even with the most intelligent software. This also applies to cloud-controlled APs, which in theory have scalable, unbounded processing power. Hence a solid Wi-Fi hardware design that ensures all the use cases can be met under all operating conditions while optimizing the system cost is essential for a good Wi-Fi product. The key hardware blocks in a typical Wi-Fi system (AP or client) are shown in Figure 5.1.

All subsystems in an AP or client platform are controlled or managed by a platform processor or host CPU. The Wi-Fi subsystem attaches to the host CPU using a digital interface that is typically PCIE, SDIO, or USB interface. The

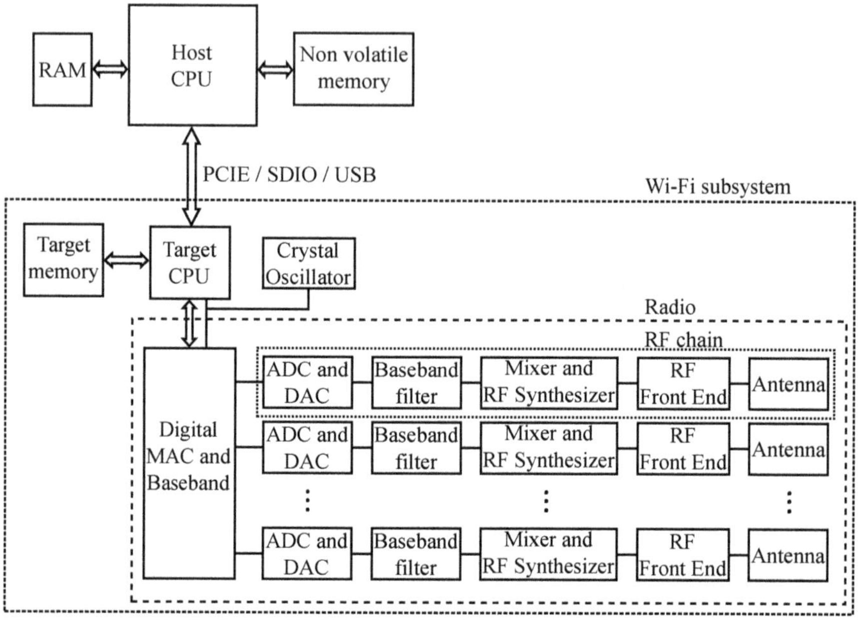


**Figure 5.1** Wi-Fi hardware block diagram.

maximum data transfer rate of this digital interface is typically rated higher than the maximum wireless physical (PHY) layer data rate of the Wi-Fi subsystem, or else it would become the bottleneck for wireless throughput. For example, the maximum throughput of a $2 \times 2$ HE80 (maximum PHY rate of 1200 Mbps) capable USB adapter supporting USB 2.0 digital interface will be limited by the USB 2.0 data rate of 480 Mbps. The Wi-Fi subsystem consists of a radio, and optionally, a target processor along with its own target memory. The target processor, if present, can potentially offload or accelerate some Wi-Fi tasks, that would otherwise run on the host CPU, resulting in a reduced host CPU utilization. The radio block is responsible for the core Wi-Fi medium access control (MAC) and PHY tasks. It typically consists of a digital implementation of MAC combined with a baseband portion of PHY and one or more RF chains. In modern wireless transceiver design, it is common to split the PHY implementation into a digital baseband portion and an analog RF portion. In simple terms, the baseband portion is a complete PHY transmitter and receiver implementation for a carrier frequency of 0 Hz. The RF portion, also referred to as RF chain, essentially acts as a frequency translator between 0 Hz and the desired channel frequency. During transmission, it transforms the digital baseband signal to an analog RF signal, and during reception, it transforms the incoming RF signal to a digital baseband signal. A MIMO product will have at least as many RF chains as the number of spatial streams (NSS) it supports and at least as many antennas as the number of RF chains. Each RF chain consists of an analog-to-digital converter (ADC), a digital-to-analog converter (DAC), a baseband filter, a mixer, an RF synthesizer, an RF front-end, and an antenna. The ADC and DAC block perform the conversion of digital signal to analog signal and vice versa. The baseband filter in the RF chain filters out any emissions outside the intended RF spectrum during transmission and also helps reject adjacent channel interference during reception. The mixer and RF synthesizer together perform the task of frequency translation. The RF front-end consists of a switch to toggle between Tx and Rx, a power amplifier (PA) and a low noise amplifier (LNA) for amplifying signals during transmission and reception, respectively. The antenna converts RF signals into electromagnetic waves and vice versa. The target processor, digital MAC and baseband block, ADC and DAC block, and RF synthesizer require a reference clock signal to generate the appropriate clock frequency for their operation. This reference clock signal is typically supplied by a crystal oscillator.

### 5.1.1 Radio Requirements

As is evident from the block diagram, the radio consists of several analog circuits whose output always includes noise and whose characteristics depend on physical parameters such as voltage, temperature, and frequency of operation. The circuits are also subject to part-to-part variations owing to the manufacturing tolerance of their components. For example, a crystal oscillator rated with a frequency stability of 20 parts per million (ppm) accuracy will cause the RF synthesizer at 5765-MHz channel frequency to have a frequency error of $\pm 5765 \cdot 20 = \pm 115.3\,\text{kHz}$. The frequency stability of a crystal oscillator also varies with temperature and could become the limiting factor for the operational temperature range of the device. A few of the common impairments in a radio implementation are listed below.

1. *Rx DC offset*: Introduced by baseband amplification stages and carrier leakage inside the mixer. Reduces the ADC dynamic range and Rx performance.

2. *IQ mismatch*: Caused by phase and gain imbalance between the in-phase and quadrature-phase baseband stages. Distorts both transmitted and received signals.

3. *Tx carrier leak*: Caused by carrier leakage inside mixer which gets amplified by the PA. Wastes transmitted energy and makes it harder for receivers to detect and decode frames.

4. *Phase noise*: Caused by crystal oscillator and the RF synthesizer. Distorts both the transmitted and received signals.

5. *PA nonlinearity* : Caused by PA nonlinearity that is characterized by the 1 dB compression point of PA. Distorts the transmitted signal. Since orthogonal frequency division multiplexing (OFDM) waveforms have high peak-to-average power ratio (PAPR), PA nonlinearity usually limits the maximum transmitted power for higher-order modulations. PA nonlinearity can also create side lobe emissions that cause adjacent channel interference.

Modern radio implementations mitigate radio impairments by performing radio calibration, that involves measuring the amount of radio impairment and compensating for it either in the digital baseband or in the analog domain. For example, Rx DC offset can be compensated for by adding a DC signal of equal magnitude but opposite polarity to the received signal before the ADC block. Similarly, PA nonlinearity can be mitigated by predistorting the signal with the PA's inverse transfer function before feeding the signal to the PA. The analog circuit design needs to have built-in provisions for such measurement and compensation. The radio calibration data containing the compensation values is measured across various parameters covering the operation range such as different channel frequencies, and Tx and Rx amplifier gains. These calibrations can either be done one time during the hardware manufacturing process or at run-time. Some radio calibrations such as PA nonlinearity have to be done at run-time because the PA transfer function can vary with temperature and aging. The IEEE 802.11 standard [1, 2] defines some radio requirements to ensure that all Wi-Fi devices deliver a minimum guaranteed PHY performance.

#### 5.1.1.1 Frequency Tolerance

The standard requires all Wi-Fi devices to derive both the symbol clock and RF carrier frequency for all antennas from the same reference oscillator. This enables receivers to estimate the symbol clock frequency offset from the carrier frequency offset and vice versa, thereby simplifying receiver implementation. The standard permits a frequency tolerance of 20 ppm in the 5-GHz band and 25 ppm in the 2.4-GHz band at the transmitter. This implies all Wi-Fi receivers should be designed to handle this frequency error in addition to any Doppler-induced frequency offset.

#### 5.1.1.2 Tx Spectral Mask

The Tx spectral mask specifies an upper bound on power spectral density for emissions that are outside the desired channel bandwidth. This is specified in dBr units (dB relative to the maximum spectral density of the transmitted signal). The Tx spectral mask requirement controls the amount of interference caused to other devices operating in nonoverlapping channel frequencies. Figure 5.2 shows the Tx spectral mask for HE PPDUs of different bandwidth. For preamble punctured transmissions, the signal leakage from the occupied subchannel into the punctured subchannel has to be below -20 dBr starting 0.5 MHz from the boundary of the punctured subchannel.

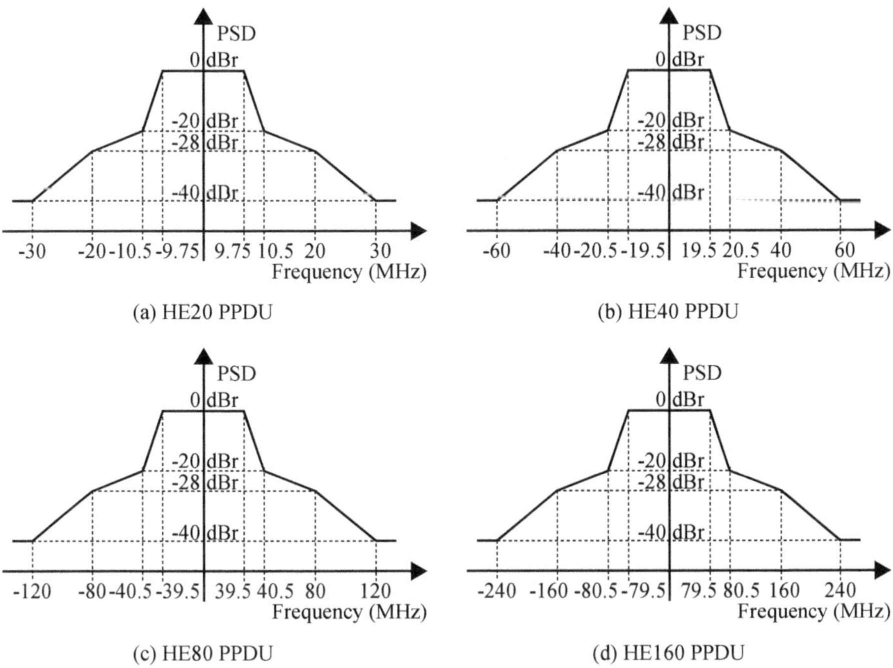


**Figure 5.2** Transmit spectral mask for different bandwidths.

#### 5.1.1.3 Tx Spectral Flatness

The Tx spectral flatness requirement specifies that the average constellation energy in each subcarrier should not deviate by more than $\pm 4$ dB from the average energy across all subcarriers. This requirement ensures that the transmitted power is uniformly distributed across all occupied subcarriers. The Tx spectral flatness requirement is verified only with BPSK modulation. Any unoccupied subcarriers are exempt from this requirement.

#### 5.1.1.4 Tx Carrier Leakage

At each antenna, the power measured at the carrier frequency using a resolution bandwidth of 78.125 KHz is required to be below MAX(P-32, -20) dBm, where P is the Tx power per antenna in dBm.

#### 5.1.1.5 Tx EVM

The amount of distortion or modulation accuracy in a Wi-Fi frame is measured by a metric called error vector magnitude (EVM). EVM encompasses all forms of distortion such as PA nonlinearity, IQ mismatch and phase noise. EVM for a signal is generally defined as the root mean square (RMS) average magnitude of the error vector normalized to the ideal signal amplitude. The error vector here refers to the vector between the actual constellation point as seen by a receiver and the ideal constellation point as shown in Figure 5.3. In the context of an OFDM frame, the RMS average of the error vector is taken across all subcarriers and OFDM symbols and the result is normalized by the average power of the OFDM frame. The measurements for at least 20 OFDM frames is then averaged to obtain the EVM. The EVM is often expressed in dB unit as $\text{EVM(dB)} = 20 \cdot \log_{10}(\text{EVM})$. An alternate way to interpret EVM is that it is simply the inverse of the SNR of the transmitted signal, where signal distortion is viewed as additive noise. Therefore, $\text{EVM(dB)} = -\text{SNR(dB)}$.

The IEEE 802.11ax standard [1] specifies a maximum EVM requirement for transmitted PPDUs depending on the MCS and type of PPDU as detailed in Table 5.1. Higher MCS requires lower EVM as the distance between constellation points shrinks with higher modulations. For example, MCS 0 SU PPDU has a Tx EVM requirement of only -5 dB, which means the SNR measured directly at the transmitter can be as low as 5 dB. The Tx EVM is dependent on the transmit power as signals with higher transmit power typically experience more nonlinearity

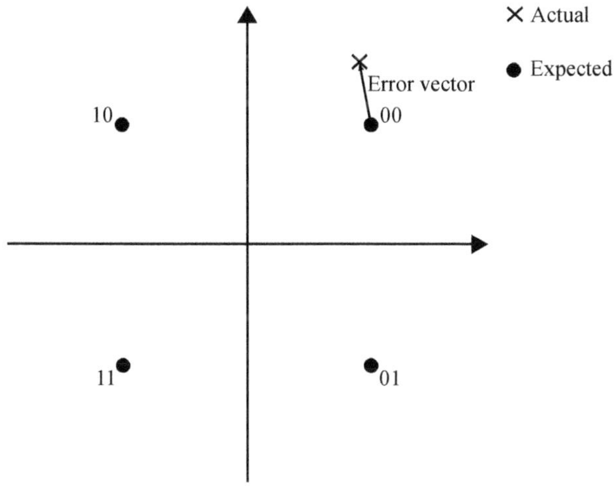


**Figure 5.3** Error vector in a constellation diagram.

in the radio transmit path. As the EVM requirement varies across different MCS, most implementations try to maximize the transmit power for each MCS. In some implementations, the difference in Tx power between MCS 11 and MCS 0 could be as high as 10 dB, which complicates the rate adaptation algorithm design as explained in more detail in Section 5.2.1. Although the standard doesn't explicitly specify any Tx EVM limits for NDP frames, the Tx EVM of NDP frames can impact the quality of channel measurements, which in turn affects TxBf and MU-MIMO performance. SU and MU beamformer implementations have to take this into consideration when deciding the Tx power for NDP frames. Finally, note that STAs have tighter EVM requirements for trigger based (TB) PPDUs depending on the Tx power.

#### 5.1.1.6 Additional Requirements for TB PPDU Transmission

TB PPDUs have certain additional requirements to mitigate synchronization and interference issues during reception at the AP. The Tx power for a TB PPDU is indirectly specified by the AP using the UL target Rx power subfield in the user info field of trigger frame. The STA is required to measure the path loss using the its Rx RSSI measurement of trigger frame and the AP Tx power subfield in the common info field of trigger frame. The calculated path loss is used to adjust the STA's Tx

**Table 5.1**
IEEE Transmit EVM Requirement

| *MCS* | *Tx EVM for SU, ER SU, MU PPDU* | *Tx EVM for TB PPDU* | |
|---|---|---|---|
| | | *Tx power>MCS 7 Tx power* | *Tx power≤MCS 7 Tx power* |
| 0 | -5 dB | -13 dB | -27 dB |
| 1 | -10 dB | -13 dB | -27 dB |
| 2 | -13 dB | -13 dB | -27 dB |
| 3 | -16 dB | -16 dB | -27 dB |
| 4 | -19 dB | -19 dB | -27 dB |
| 5 | -22 dB | -22 dB | -27 dB |
| 6 | -25 dB | -25 dB | -27 dB |
| 7 | -27 dB | -27 dB | -27 dB |
| 8 | -30 dB | -30 dB | -30 dB |
| 9 | -32 dB | -32 dB | -32 dB |
| 10 | -35 dB | -35 dB | -35 dB |
| 11 | -35 dB | -35 dB | -35 dB |

power of TB PPDU to meet the specified UL target Rx power. The standard requires all STAs to support a minimum TB PPDU Tx power of MAX(P-32, -10) dBm where P is the maximum Tx power supported by the client.

The standard also provides a provision for a STA to indicate its Tx power accuracy and Rx RSSI measurement accuracy to the AP using the device class subfield of HE PHY capability IE. The AP can factor this accuracy in its uplink Tx power and uplink rate adaptation algorithms. Table 5.2 shows the minimum expected Tx power accuracy and Rx RSSI measurement accuracy for class A and class B STAs.

**Table 5.2**
Power Accuracy for Class A and Class B STAs

| *Parameter* | *Class A* | *Class B* |
|---|---|---|
| Absolute Tx power accuracy | ±3 dB | ±9 dB |
| Relative Tx power accuracy | Not applicable | ±3 dB |
| Rx RSSI measurement accuracy | ±3 dB | ±5 dB |

A STA is required to measure the carrier frequency offset (CFO) error using the trigger frame and then compensate the CFO error and symbol clock error on the TB PPDU transmission. The residual CFO error on the TB PPDU transmission measured at the 10% point of cumulative distribution function (CDF) is required to be below 350 Hz. Such tight requirements, on the CFO error mitigates intercarrier interference during UL MU reception. Finally, the transmission start time of TB PPDU or CTS frame in response to a trigger frame should be within $16 \pm 0.4\,\mu s$ of the trigger frame ending.

#### 5.1.1.7 Rx Sensitivity

The standard requires all devices to meet a minimum Rx sensitivity as specified in Table 5.3 for different MCS and bandwidth. The packet error rate (PER) at the Rx sensitivity level should be below 10% for a PSDU length of 4096 bytes. The coding scheme for the Rx sensitivity requirement is assumed to be binary convolutional code (BCC) wherever allowed and low density parity check (LDPC) code for cases where BCC is not allowed. In an implementation, the Rx sensitivity for low MCS depends on the noise figure of the receiver and maximum gain of the receiver, whereas, at high MCS, the quantization noise at ADC, phase noise, IQ mismatch, and Rx nonlinearity also affect Rx sensitivity.

**Table 5.3**

Minimum Receive Sensitivity for Different MCS and Bandwidth

| *MCS* | *Minimum Rx Sensitivity in dBm* | | | |
|---|---|---|---|---|
| | *20 MHz* | *40 MHz* | *80 MHz* | *160 MHz* |
| 0 | -82 | -79 | -76 | -73 |
| 1 | -79 | -76 | -73 | -70 |
| 2 | -77 | -74 | -71 | -68 |
| 3 | -74 | -71 | -68 | -65 |
| 4 | -70 | -67 | -64 | -61 |
| 5 | -66 | -63 | -60 | -57 |
| 6 | -65 | -62 | -59 | -56 |
| 7 | -64 | -61 | -58 | -55 |
| 8 | -59 | -56 | -53 | -50 |
| 9 | -57 | -54 | -51 | -48 |
| 10 | -54 | -51 | -48 | -45 |
| 11 | -52 | -49 | -46 | -43 |

#### 5.1.1.8 Maximum Rx Input Level

The maximum Rx input level requirement specifies that all devices should provide a PER of less than 10% across all supported PHY rates for a PSDU length of 4096 bytes at a Rx input level of -20 dBm in 2.4 GHz and -30 dBm in 5 GHz. Essentially this requirement along with the Rx sensitivity requirement specifies a dynamic range requirement for the receiver. In order to meet this dynamic range requirement, there should be adequate programmable Rx gain or attenuation in the automatic gain control circuitry of the receiver.

#### 5.1.1.9 Rx Blocker Rejection

Rx blocker rejection signifies the amount of tolerance of a receiver to an interfering Wi-Fi signal on a different channel. The IEEE 802.11ax standard specifies a minimum blocker rejection requirement for both adjacent and nonadjacent channels across different MCS. Assuming the bandwidth of desired channel is B MHz, an adjacent channel is defined as a channel with center frequency B MHz away from the center frequency of desired channel. A nonadjacent channel is defined as any channel with center frequency at least $2 \cdot B$ MHz away from the center frequency of desired channel. Blocker rejection is measured as the minimum difference in power levels between the blocker signal and the desired signal at which the PER becomes 10% while the desired signal level is 3 dB higher than the minimum Rx sensitivity specified in Table 5.3. For example, if a receiver has adjacent channel rejection of 10 dB for MCS 4, 80 MHz bandwidth, it implies the PER is less than or equal to 10% when the desired signal level is -61 dBm and the adjacent channel Wi-Fi signal level is -51 dBm. Table 5.4 shows the adjacent and nonadjacent channel rejection requirements for different MCS.

### 5.1.2 Host Processor

There are several factors that determine the choice of a host processor and in many use cases Wi-Fi may not be the deciding factor. For example, the host processor in a smart phone is determined by several non-Wi-Fi requirements. One of the key metrics to consider in general though is the amount of spare host CPU utilization (in percentage) required for non-Wi-Fi applications when the Wi-Fi subsystem is running at peak throughput and traffic load. This in turn influences the choice of processor architecture and the CPU speed. For instance, 50% spare host CPU utilization might be adequate for an access point while a smart phone might require

**Table 5.4**

Blocker Rejection Levels for Different MCS

| *MCS* | *Adjacent Channel Rejection (dB) for 20-,40-,80-,160-MHz Channel* | *Nonadjacent Channel Rejection (dB) for 20-,40-,80-,160-MHz Channel* |
|---|---|---|
| 0 | 16 | 32 |
| 1 | 13 | 29 |
| 2 | 11 | 27 |
| 3 | 8 | 24 |
| 4 | 4 | 20 |
| 5 | 0 | 16 |
| 6 | -1 | 15 |
| 7 | -2 | 14 |
| 8 | -7 | 9 |
| 9 | -9 | 7 |
| 10 | -12 | 4 |
| 11 | -14 | 2 |

more than 90% spare host CPU utilization. Another metric to consider is the number of available CPU cores or threads to enable parallel processing. This can be useful to minimize packet processing latency especially when there are many traffic flows. Finally, in battery powered devices, the power consumption of the host processor is another important metric.

### 5.1.3 Memory Size

Memory size is determined by operating system requirements, application requirements, and Wi-Fi packet buffering requirements. The Wi-Fi packet buffering requirement differs significantly between AP and client. Inadequate buffer allocation for Wi-Fi can result in frequent packet drops and poor performance. TCP traffic flows are most affected by such packet drops owing to the TCP congestion control mechanism, that throttles the traffic rate in response to any packet drop. A client needs to buffer packets during power save, off-channel scan, roaming, medium contention, and certain time durations of high PER. Also, whenever there is a momentary speed mismatch between the application layer and the wireless PHY rate, packets have to be buffered in both downlink and uplink directions, to avoid packet drops. In an AP, the Wi-Fi packet buffering requirement increases with the number of clients being served. In addition to data packets, management packets also consume a significant number of buffers on an AP, that depends on both the number of SSIDs and client count. To approximately estimate the number of required buffers for downlink traffic, the average time taken by a packet from the time it arrived at the backhaul link to the time it is transmitted on air can be multiplied by the backhaul link speed. For example, if the backhaul link is a 1-Gbps Ethernet link and if the average time a packet stays on the AP is 0.1s, then the approximate buffer size required for downlink traffic is $1000 \cdot 0.1 = 100$ Megabits. Similarly in the uplink direction, the average time required to serve a received packet can be multiplied by the maximum wireless PHY rate of a connected client to arrive at an approximate receive buffer size. Finally, a certain amount of buffers have to be dedicated for clients in the power save state, multicast traffic, and management packets. The total memory size has to be chosen considering all these important aspects. The memory type commonly used in most implementations is double data rate (DDR) synchronous dynamic random-access memory (SDRAM), which is a volatile memory, (i.e., one whose data contents are lost without power). It is also common for a Wi-Fi system to have an additional nonvolatile memory to store sensitive boot code and radio calibration data obtained at manufacturing time as these need to be retained even without power.

## 5.2 SOFTWARE

Software algorithms are essential to adding value to any Wi-Fi product and bringing out the strengths of a good Wi-Fi hardware design. Although implementing everything using an application specific integrated circuit (ASIC) hardware might be cost optimal, retaining some of the important logic in software provides the ability to make improvements over time and continuously enhance the user experience. Moreover, such flexibility is necessary to address a broad range of applications as well as issues that may be uncovered over the life cycle of a Wi-Fi product. In some cases, the same Wi-Fi ASIC is used in both AP and STA implementations although the key requirements on AP and STA are very different. The challenges in Wi-Fi software algorithm design are highly implementation specific so this section only touches upon some of the common software algorithm design challenges in AP and STA.

### 5.2.1 Rate Adaptation

The rate adaptation algorithm is responsible for determining the NSS, bandwidth (BW), modulation and coding scheme (MCS), and guard interval (GI) for all transmissions. In the case of a Wi-Fi 6 AP, there is an additional uplink rate adaptation algorithm meant for triggered uplink transmissions. Assuming a $2 \times 2$ HE80 capable client, there are two choices for spatial streams, three choices for BW, twelve choices for MCS and three choices for GI, which results in a total of $2 \cdot 3 \cdot 12 \cdot 3 = 216$ different PHY rate choices. Each of these 216 PHY rate choices have a corresponding PER depending on the link budget, channel condition, and the rate adaption algorithm needs to find the PHY rate choice that optimizes the throughput. There are two key metrics to evaluate a rate adaptation algorithm:

1. Time taken to converge to the optimal PHY rate choice;

2. The percentage of transmit data airtime that employs the optimal PHY rate.

The first metric captures the reaction time to a sudden change in path loss or channel condition. This metric needs to be as small as possible and it is a measure of how well the algorithm performs whenever there are variations due to the movement of the devices or the surroundings. The second metric reflects the steady state performance of the algorithm and is a measure of how close it is to an ideal algorithm. A good rate adaptation algorithm is one that minimizes the first metric and maximizes the second metric. In practice though, it is a delicate trade-off to balance the two metrics depending on the use case. One of the big challenges in Wi-Fi rate adaptation is to obtain throughput estimates for different PHY rates which is possible only by attempting few PPDUs at different PHY rates. The only feedback available from the receiver of a PPDU is the BA from which the PER can be estimated within some degree of uncertainty. The transmitter of a PPDU is hence unaware of the Rx RSSI observed by the receiver, and similarly, the receiver of a PPDU is unaware of the transmitter's Tx power. Also, note that PER can result from either poor Rx SNR or collisions and distinguishing between these two is generally a challenge. If the PER is due to collision, simply retrying at the same PHY rate or protecting the transmission with RTS-CTS is a better solution than reducing the PHY rate. Periodically attempting all the PHY rates incurs significant overhead so some intelligence is required to select a few PHY rate candidates to attempt and then predict bounds on the throughput estimate for other PHY rates. For example, if a PER of 10% is encountered with 1 spatial stream, HE80, MCS 4 PHY rate, then with high probability the PER is greater than 10% for 1 spatial stream, HE80, MCS 5 to 11 PHY rates and lower than 10% for 1 spatial stream, HE80, MCS 0 to 3 PHY rates. Further adding to the complexity, the Tx power and Tx EVM can vary with MCS, spatial streams, and BW. These practical aspects make the throughput prediction for untried PHY rates difficult. In some cases, the channel feedback can serve as a guide to find the optimal spatial streams or MCS.

In the case of triggered uplink transmission, the AP specifies the target Rx RSSI for the TB PPDU but the AP is unaware of the STA's maximum Tx power limit. The AP has to assess the actual path loss, approximately predict the optimal PHY rate for TB PPDU, and then make fine adjustments based on the actual throughput. One additional challenge in uplink rate adaptation is that the PER and hence optimal PHY rate also depends on the RU allocation since the Rx SNR per subcarrier increases with decreasing RU size. It is a complex task for an AP to adjust the PER estimate or throughput estimate according to RU size because the AP is unaware of the STA's maximum power spectral density limit and the STA's Tx power limit.

### 5.2.2 Scheduler

The scheduler is responsible for determining the order of transmitting PPDUs. The Wi-Fi standard defines EDCA protocol, that describes how medium access is prioritized for QoS data frames among different TIDs. Wi-Fi implementations maintain separate queues for different TIDs and management traffic. In the case of an AP, there are additional queues maintained per STA per TID, a power save queue meant for traffic destined to STAs in the power save state, and another queue for multicast and broadcast traffic. As packets arrive, they are sent to the appropriate queues where they are buffered, and the scheduler decides the order of transmitting frames from the different queues. Upon completion of a transmission, the successfully acknowledged MSDUs are removed from the corresponding queue based on the BA and the scheduler determines whether the next PPDU transmission should be a retransmission attempt or a fresh PPDU transmission for a different queue. The scheduler is typically implemented in software to allow flexibility in addressing any exceptions. The scheduler algorithm on an AP is much more complex than the one on a STA. The goal of the scheduler is to minimize queuing delay of traffic while ensuring fairness in serving different types of traffic and different STAs. Sometimes the fairness aspect is dependent on the specific use case. Establishing a scheduling priority when there is resource contention across unicast traffic, multicast traffic, broadcast traffic, and management traffic is a critical task and sometimes trade-offs have to be made depending on the use case. For example, if an AP with unicast downlink traffic to 100 connected STAs is suddenly bombarded with probe requests from 50 nonassociated STAs, it will experience a conflict between serving connected STAs and allowing new connections. If the AP decides to prioritize unicast traffic, the probe requests won't be responded to in time and the 50 STAs that are scanning, would move on to another scan channel resulting in a significant delay to get connected. If the AP prioritizes serving the probe requests, it can affect the latency or user experience for the 100 connected STAs. This scenario can practically occur in AP deployments at public places such as train stations or bus stops where there are numerous clients in transit. A scheduler needs to maintain a fair balance in airtime allocation across different types of traffic and enforce upper bounds on airtime allocation to each traffic type in a given time window.

One question to think about is, how should fairness be measured? Should fairness be measured in terms of number of packets scheduled or airtime allocated in a certain time window? Since different clients have different capabilities and are at different distances from the AP, the PHY rate to each client is different. Moreover, Wi-Fi 1 to 3 generations do not allow aggregation while Wi-Fi 4 and above use aggregation. So there is significant disparity in PPDU airtime between different Wi-Fi generations. For these reasons, it is more appropriate to measure fairness in airtime and several modern Wi-Fi implementations have already transitioned to this metric. With multi-TID AMPDU, OFDMA, and TWT in Wi-Fi 6, there are more dimensions of flexibility available to the scheduler. A Wi-Fi 6 AP has to perform scheduling for uplink traffic in addition to downlink traffic and ensure a fair airtime balance between downlink and uplink traffic. With MU-MIMO and MU-OFDMA in Wi-Fi 6, fairness in resource allocation has to be measured and tracked across time, frequency, and spatial streams instead of only airtime. Finally, another challenging aspect specific to a Wi-Fi 6 AP scheduler is determining when to do SU, MU-MIMO, MU-OFDMA, and deciding the group of users in MU PPDU or TB PPDU. A good scheduling algorithm is critical to minimizing latency, maximizing efficiency, and ensuring fairness in a Wi-Fi network.

### 5.2.3 Buffer management

Buffer management is a critical but often overlooked aspect in Wi-Fi implementations. Any Wi-Fi device has a finite amount of memory buffers, determined by the RAM size and memory allocated for Wi-Fi subsystem to hold packets. These buffers have to be allocated across different queues, and packets will be dropped whenever there is insufficient buffer allocation. Buffer management is a more challenging problem for an AP compared to a STA. A static buffer allocation for different queues will result in poor usage of the buffers compromising the overall performance as the incoming traffic distribution to different clients and TIDs is not identical and varies with time. Hence, many implementations opt for a dynamic buffer allocation. However, in dynamic buffer allocation, proper upper bounds on queue size have to be enforced for each queue; otherwise, one queue could consume a vast majority of buffers thereby starving the other queues. A common problem is the fast client, slow client scenario, where there are two STAs (STA1 and STA2) connected to an AP and each STA has a UDP downlink traffic stream of 10 Mbps each. Assume STA1 has a downlink PHY rate of 100 Mbps and STA2 initially had a downlink PHY rate of 24 Mbps but temporarily dropped to 6 Mbps. The queue size of STA2 will build up indefinitely owing to its PHY rate being lower than the incoming traffic speed. If there is no upper limit enforced on the buffer allocation for STA2, then STA1 will be starved of buffers, resulting in packet drops for STA1 even though STA1 has an excellent wireless connection. Similarly, appropriate upper limits have to be enforced for the power save queue and multicast, broadcast, and management traffic queues. The buffer allocation strategy has to take into consideration the most common use case scenarios of the product and try to match the incoming traffic type distribution.

### 5.2.4 Packet Processing Speed

As the complexity of scheduling and rate adaption algorithm increases, it is important to pay attention to the packet processing speed as well. If the complexity of an algorithm is high relative to the available CPU processing power, the CPU time spent on running the algorithm might offset the supposed benefits of the algorithm and limit the peak performance. The peak throughput of a Wi-Fi system is determined by the packet processing speed as well as the maximum wireless PHY rate. A common test case to measure the packet processing speed of a Wi-Fi system is to run UDP traffic at maximum wireless PHY rate with 64 byte size packets and then measure the maximum number of packets transmitted per second. At such small packet size, the peak performance is limited by the packet processing speed rather than the wireless PHY rate. The packet processing speed is a useful metric to determine an upper bound on the peak performance of the system. For example, if the packet processing speed of a $2 \times 2$ HE80 capable device is measured to be 10,000 packets per second (pps), then the peak throughput with 1,500 byte size packets will be limited to $10,000 * 1,500 * 8 = 120$ Mbps even though the maximum wireless PHY rate capability of the device is 1,200 Mbps. So, it is important to maximize the packet processing speed and adapt algorithm complexity according to the available CPU processing power. It is worth noting that the packet processing speed may also be limited by the hardware digital interface between the host CPU and Wi-Fi subsystem.

### 5.2.5 Channel Selection

Since Wi-Fi operates in unlicensed spectrum, it is important for an AP to periodically scan the spectrum and choose an operating channel with minimum interference in order to provide a reliable Wi-Fi service. The interference source could be either a Wi-Fi device or a non-Wi-Fi device such as Bluetooth, microwave, cordless phone, or baby monitor. For this reason, some Wi-Fi implementations integrate a spectrogram functionality or a dedicated scanning radio to quantify the impact of all interference sources across the entire spectrum. Below four aspects need to be considered while evaluating the impact of non Wi-Fi interference sources:

1. Interference level as seen by AP;
2. Frequency range of interference;
3. Duty cycle: percentage of time interference is active;
4. Number of false Wi-Fi frame detections caused by interference.

Some interference sources are wide-band, impacting multiple Wi-Fi channels whereas some are narrow-band, affecting only one channel. Certain interference sources, although weak, can resemble some attributes of a Wi-Fi preamble resulting in false Wi-Fi frame detections in certain Wi-Fi implementations. A high false Wi-Fi frame detection rate can cause an AP to miss detection of actual Wi-Fi frames and delay transmission of time-sensitive frames such as beacons. So, if a channel has an excessive false Wi-Fi frame detection rate, an AP should try to avoid it. When evaluating Wi-Fi interference, a channel selection algorithm should take into consideration the number of neighboring APs operating in different channels along with each of their operating bandwidth. The duty cycle of a Wi-Fi interferer is highly dynamic and changes significantly with time, so optimizing for the duty cycle can be a secondary metric to consider after first minimizing the number of cochannel neighbors. A channel selection algorithm should also take a balanced view between picking the optimal channel and minimizing the number of channel changes. Frequent channel changes can be disruptive in certain use cases especially to connected clients that don't support the 802.11h channel switch announcement. For this reason, some channel selection algorithm implementations confine channel changes to time durations when there is no active traffic, such as during nighttime.

The channel selection problem is more challenging in multi-AP deployments where the self-interference of the network is more than external interference sources. Sometimes, multiple independent Wi-Fi networks are deployed in the same venue, making channel selection more complicated. For example, different floors in a multistory office space may be leased to different companies resulting in independent Wi-Fi networks at each floor. In a multi-AP Wi-Fi network, if each AP decides its channel independently it can sometimes result in a channel plan that changes continuously. Let us consider an example scenario where there are only two available channels 1, 11, and two APs (AP1, AP2) that are deployed within hearing range of each other. Let us assume both APs boot up around the same time and happen to start with channel 1. After some time both APs learn that channel 11 has no cochannel neighbors and hence switches to channel 11. This action then triggers both APs to change channel once again and this cycle continues forever. To avoid this type of problem, enterprise Wi-Fi deployments typically have a centrally managed channel allocation algorithm that takes into account the neighbor list for each AP and arrives at a channel plan that optimizes for the least number of cochannel neighbors.

### 5.2.6 STA Roaming Algorithm

When there are multiple APs in the vicinity of a STA, the STA roaming algorithm decides the AP to which a connection attempt is made. This decision happens during the initial connection and is periodically revisited to ensure that the STA is always connected to the best possible AP. Since a STA can be mobile, the roaming algorithm plays a crucial role in maintaining a good wireless link especially in a multiple AP deployment. Multiple AP in this context can be a single physical AP supporting concurrent 2.4-GHz and 5-GHz operation or multiple VAPs. Nowadays, large homes and homes with brick walls are deploying multiple APs to provide whole-home coverage. Although an AP can perform neighbor discovery and use 802.11v BTM to steer clients to a better AP or a different frequency band, the final decision is taken by the STA roaming algorithm. A roaming algorithm considers various metrics, such as beacon receive signal strength indicator (RSSI) from its own off-channel scan report, AP's capabilities, number of APs in a channel, 802.11k neighbor report, 802.11v BTM request from APs, and medium access time in different channels, before deciding the frequency band and AP to establish connection. The roaming algorithm also decides how often to perform off-channel scan and when to start aggressively scanning for a better AP. If a STA is moving away from an AP fast, it needs to initiate and complete roaming to another nearby AP before the STA completely goes out of range from the current AP. So a roaming algorithm periodically monitors the signal strength and quality metrics of the wireless link, such as beacon misses, PHY rate, PER, and RSSI, and triggers a roaming attempt if the link quality degrades. Frequent roaming can also be disruptive to users, especially if 802.11r is not supported, so a roaming algorithm should take a balanced approach.

### 5.2.7 STA Power Save Algorithm

One of the most important Wi-Fi software algorithms on a battery-powered STA is the power save algorithm, that decides when to enter and exit power save (PS) state, and specifies what type of PS mechanism (PS-Poll, UAPSD, TWT, WNM sleep, etc.) to use. The challenge for a PS algorithm is to balance performance and power consumption depending on the use case. If a STA frequently goes to PS state, it will experience increased latency for downlink traffic and the STA may not get a fair share of the downlink airtime. This is especially true for TCP traffic, which is sensitive to the round trip of time of TCP segments. A STA has better knowledge of its uplink traffic compared to downlink traffic. It is therefore easier for a STA to optimize its power consumption when the traffic is dominated by uplink traffic. The challenge in assessing the downlink traffic arises because the AP does not inform a STA in PS state about the amount of buffered downlink traffic and only indicates whether there is buffered traffic in the TIM IE. This makes it difficult for a STA to decide the right PS scheme. For example, if a STA is aware that there is only one packet buffered at AP, then the STA can use the PS-poll scheme but if there are too many buffered packets, the STA can simply exit the PS state. As another example, if the STA becomes aware of an ongoing voice call, it would appropriately use TWT. Certain battery-powered IoT use cases might opt for WNM sleep scheme to maximize the sleep duration. Knowledge of the traffic pattern or the actual application that is driving the traffic can significantly aid a PS algorithm in choosing the right PS strategy.

### 5.2.8 Capability Advertisements

The capability advertisements in beacons, association request, and response frames are typically handled in software. It is important for software to advertise the right parameter values matching the actual capabilities of the hardware. Certain incorrect parameter values can result in performance or connectivity issues. In particular, hardware parameters such as minimum MPDU start spacing, HE packet extension, and supported MCS, NSS, BW, GI have to be advertised correctly to avoid abnormal packet drop events.

## 5.3 REGULATORY REQUIREMENTS

In addition to IEEE specified requirements, there are some country- or region-specific requirements known as regulatory requirements. In order to operate a Wi-Fi device in a particular region, it has to comply with the local regulatory requirements. The regulatory requirements specify the channel frequencies allowed for Wi-Fi operation and the various conditions under which Wi-Fi can operate in a region. Since there are many countries and country-specific requirements, the following sections describe the main requirements for only United States and European Union (EU) regions. The regulatory rules are specified by the Federal Communications Commission (FCC) in the United States and by European Telecommunications Standards Institute (ETSI) in the EU.

### 5.3.1 Allowed Channel Frequencies

The 2.4-GHz band (2400 to 2483.5 MHz) has a total of 14 overlapping channels, numbered 1-14 as illustrated in Figure 5.4. The center frequencies of each channel are separated by 5 MHz except for a 12-MHz space before channel 14. Except for channel 14, the channel center frequency for all other channels in 2.4-GHz band can be derived from the channel number using (5.1).

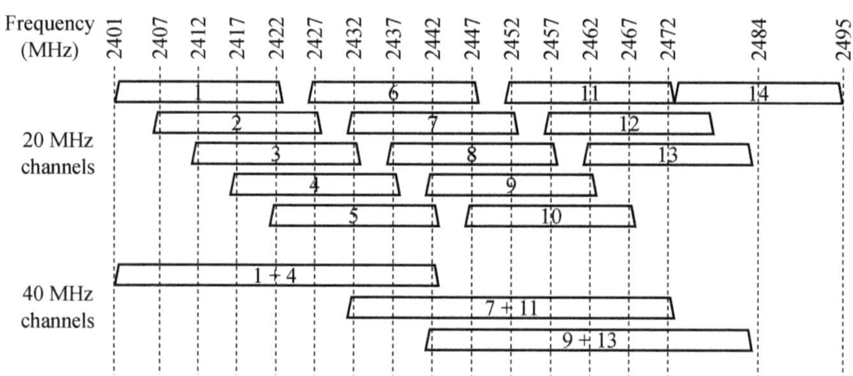


**Figure 5.4** 2.4-GHz channels.

$$\text{Channel center frequency (MHz)} = 2407 + 5 \cdot \text{Channel number} \tag{5.1}$$

The width of each channel is 22 MHz if 802.11b is supported and 20 MHz if 802.11b is not supported. Two adjacent channels can be bonded to form a 40-MHz channel in 2.4-GHz band but this is used mainly in single AP deployments. Channels 1 to 11 are allowed for Wi-Fi operation in the United States [3] whereas channels 1 to 13 are allowed in the EU [4] and all channels 1 to 14 are permitted in Japan. In the United States, only three nonoverlapping 20-MHz channels are possible in the 2.4-GHz band.

The 5-GHz spectrum in the United States [5] has a total of 24 nonoverlapping 20-MHz channels, numbered 36 to 165 and is divided into four Unlicensed National Information Infrastructure (U-NII) bands as follows:

1. U-NII-1 comprises four nonoverlapping 20-MHz channels, 36-48 in the 5150- to 5250-MHz band.
2. U-NII-2A comprises four nonoverlapping 20-MHz channels, 52-64 in the 5250- to 5350-MHz band.
3. U-NII-2C comprises eleven nonoverlapping 20-MHz channels, 100-140 in the 5470- to 5725-MHz band.
4. U-NII-3 comprises five nonoverlapping 20-MHz channels, 149-165 in the 5725- to 5850-MHz band.

The channels in U-NII-3 band are not allowed for Wi-Fi operation in the EU [6]. Figure 5.5 shows the allowed 5-GHz channels for 20-, 40-, 80- and 160-MHz bandwidth. In the 5-GHz band, the channel center frequency can be obtained from the channel number using (5.2).

$$\text{Channel center frequency (MHz)} = 5000 + 5 \cdot \text{Channel number} \tag{5.2}$$

### 5.3.2 In-Band Power and Out-of-Band Emission Limits

The FCC and ETSI specify limits for maximum in-band power and out-of-band emissions as summarized in Tables 5.5 and 5.6. FCC specifies the in-band power limits in terms of maximum conducted power summed across all Tx chains and an additional power spectral density (PSD) limit usually specified in dBm/MHz unit. FCC also allows a directional gain of up to 6 dBi without any penalty with some exceptions for point-to-point systems. The directional gain is defined as the sum of physical antenna element gain and the coherent combining gain due to

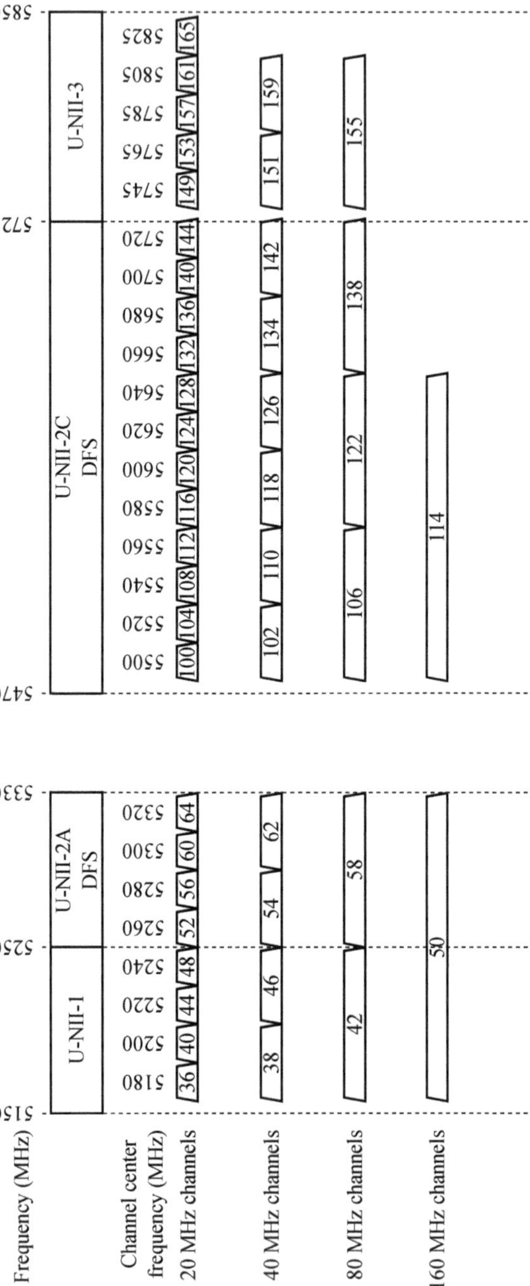


**Figure 5.5** 5-GHz channels.

**Table 5.5**
FCC In-Band Power and Out-of-Band Emission Limits

| *Frequency Range (MHz)* | *In-Band Power Limit* | *Out-of-Band Emission Limit* |
|---|---|---|
| 5150 to 5250 | For AP, the total conducted power across all chains <= 30 dBm, PSD<= 17 dBm/MHz if directional gain <= 6 dBi. If the directional gain exceeds 6 dBi, the Tx power and PSD must be reduced by the amount it exceeds 6 dBi.<br>For client, total conducted power across all chains <= 24 dBm, PSD<= 11 dBm/MHz if directional gain <= 6 dBi.<br>For point-to-point system, directional gain upto 23 dBi is allowed without reduction in Tx power. | EIRP <= −27 dBm/MHz outside 5150 to 5350 MHz. |
| 5250 to 5350 | Total conducted power across all chains <= 24 dBm, PSD<= 11 dBm/MHz if directional gain <= 6 dBi. | EIRP <= −27 dBm/MHz outside 5150 to 5350 MHz. |
| 5470 to 5725 | Total conducted power across all chains <= 24 dBm, PSD<= 11 dBm/MHz if directional gain <= 6 dBi. | EIRP <= −27 dBm/MHz outside 5470 to 5725 MHz. |
| 5725 to 5850 | For AP and client, total conducted power across all chains <= 30 dBm, PSD<= 30 dBm/500 kHz if directional gain <= 6 dBi.<br><br>For point-to-point system, there is no limit on directional gain. | EIRP <= 15.6 dBm/MHz at 5 MHz from bandedge, <= 10 dBm/MHz at 25 MHz from bandedge and <= −27 dBm/MHz at 75 MHz from bandedge. |
| 2400 to 2473 | For AP and client, total conducted power across all chains <= 30 dBm, PSD<= 8 dBm/3 kHz if directional gain <= 6 dBi.<br><br>For point-to-point system, if directional gain exceeds 6 dBi, the Tx power and PSD should be reduced by 1 dB for every 3 dB excess over 6 dBi. | EIRP measured in 100 kHz bandwidth outside 2400 to 2483.5 MHz should be 30 dB below the in-band EIRP PSD measurement in 100 kHz bandwidth. |

**Table 5.6**
ETSI In-band Power and Out-of-Band Emission Limits

| *Frequency Range (MHz)* | *In-band Power Limit* | *Out-of-Band Emission Limit* |
|---|---|---|
| 5150 to 5350 | EIRP<= 23 dBm, PSD<= 10 dBm/MHz. | EIRP <= −30 dBm/MHz in 1000 to 5150 MHz, 5350 to 5470 MHz, 5725 to 26000 MHz. |
| 5470 to 5725 | For AP, EIRP<= 30 dBm and PSD<= 17 dBm/MHz.<br><br>For client, EIRP<= 23 dBm and PSD<= 10 dBm/MHz. | EIRP <= −30 dBm/MHz in 1000 to 5150 MHz, 5350 to 5470 MHz, 5725 to 26000 MHz. |
| 2400 to 2483.5 | EIRP<= 20 dBm, PSD<= 10 dBm/MHz. | EIRP <= −10 dBm/MHz within 20 MHz from bandedge and <= −20 dBm/MHz in 20 to 40 MHz from bandedge and <= −30 dBm/MHz beyond that. |

multiple Tx chains. For example, if a TxBf capable device has 2 dBi antenna gain and 4 Tx chains, then the directional gain for 1 spatial stream transmission is $10 \cdot \log_{10}(4) + 2 = 8$ dBi. ETSI specifies the in-band Tx power limit in terms of equivalent isotropic radiated power (EIRP), which already factors in the directional gain. In some cases, AP and client have different limits.

The out-of-band emission limit is typically specified as a PSD limit measured at out-of-band frequencies and this regulation ensures Wi-Fi doesn't cause interference in non-Wi-Fi frequency bands. The out-of-band emission limit is often more stringent than the IEEE spectral mask requirement and is often the limiting factor for Tx power on bandedge channels, (i.e., channels that are on the extreme left or right boundary of allowed Wi-Fi frequency range). summarize the in-band power and out-of-band emission limits for FCC and ETSI, respectively.

### 5.3.3 Dynamic Frequency Selection

The channels in U-NII-2A and U-NII-2C band are primarily allocated for radar operation in both the United States and EU. Wi-Fi devices are allowed to operate in these channels only if there is no radar presence detected using the dynamic frequency selection (DFS) functionality. The channels requiring DFS functionality are referred to as DFS channels. The DFS functionality is enabled by the 802.11h standard and the exact requirements for DFS in the United States and EU are governed by FCC [7] and ETSI [6], respectively.

#### 5.3.3.1 Radar Characteristics

Radar is a pattern of narrow band pulse sequence used in weather, military, and satellite applications. A pulse pattern is characterized by pulsewidth or duration of pulse, pulse repetition interval (PRI), and number of pulses in a sequence. Each country specifies a number of radar types based on the pulse pattern parameters. For example, FCC radar type 3 is defined by 6-10 $\mu$s pulse width, 200-500 $\mu$s PRI, and 16-18 number of pulses. Some radar types involve chirp pulses where the center frequency of each pulse linearly increases with time and is defined by a start frequency and end frequency. Detecting radar pulses requires special signal processing capability in the hardware along with pulse pattern correlation and the challenging part is distinguishing radar pulses from Wi-Fi frames or other interference sources. One significant difference between a radar pulse and a Wi-Fi frame is that radar pulses are narrow-band whereas Wi-Fi frames have a bandwidth of at least 20 MHz. Radar pulses vary significantly in pulsewidth depending on the radar type ranging from as low as 0.5 $\mu$s to as high as 100 $\mu$s. For short pulse widths, it is generally a challenge for the AGC logic in a digital receiver to converge to the right Rx gain quickly, resulting in either signal compression or signal clipping at the ADC. A good radar detector should maximize radar detection probability while minimizing false radar detection.

#### 5.3.3.2 DFS Operation

DFS operation involves at least one master device and one or more slave devices. The AP performs the role of master and clients typically take on the slave role. Before beginning any transmission on a DFS channel, the AP should first scan the channel for radar presence continuously for a time duration of channel availability check (CAC) time. If there is no radar presence detected during the CAC time, the AP can start beaconing and begin Wi-Fi service on the channel. If a radar matching any of the specified radar types was detected during CAC time, the channel is moved to nonoccupancy list (NOL) and this channel is considered not available for a period called nonoccupancy period. Slaves are not supposed to transmit anything (including probe request) on a DFS channel until they find an AP beaconing on the channel. After an AP starts service in a channel, it is expected to perform in-service monitoring and continuously look for radar detection while Wi-Fi service is ongoing. When there are active Wi-Fi transmissions, the AP can miss detection of some radar pulses, so the DFS requirements call for a lower radar detection probability during in-service monitoring. If radar is detected during in-service monitoring, the AP should stop its transmissions, and use the 802.11h channel switch announcement frame to inform all STAs to stop transmission immediately and move to the new channel. The AP and all connected clients should vacate the channel within a time duration called channel move time and the channel is moved to the NOL list for nonoccupancy period. The aggregate of all Tx during the channel move time is called channel closing Tx time and the DFS requirement specifies this time limit. Once the nonoccupancy period for a channel expires, it can be removed from the NOL list, but an AP needs to redo the CAC before starting operation on that channel.

Since every time an AP moves to a DFS channel there is Wi-Fi service disruption for CAC time duration; ETSI allows off-channel CAC as an option to avoid this problem. Using off-channel CAC, an AP can continue service in one channel while it scans for radar presence on a different channel for noncontinuous small chunks of time. If the accumulated radar scan time off-channel exceeds the off-channel CAC time, the channel can be considered clear of radar and at that point AP can issue a 802.11h CSA to move connected clients to that channel, totally avoiding the CAC time. For slave devices used in outdoor point-to-point systems, ETSI additionally mandates DFS radar detection capability. If the slave detects radar, it should notify the master using 802.11h and the master will initiate a channel change.

Table 5.7 summarizes the various DFS requirements in both FCC and ETSI. While complying to DFS requirements is laborious and increases product certification time, it does open up a significant number of channels in the 5-GHz band. Especially in the EU, the channels that allow the highest in-band Tx power are DFS channels. Finally, Wi-Fi implementations supporting DFS channels should watch out for false radar detections and keep it under control, or else it can be disruptive to Wi-Fi service.

## 5.4 WI-FI CERTIFICATIONS

Wi-Fi certifications, although not mandatory for selling a product, add significant market value to a product. Consumers feel assured that a Wi-Fi certified product has undergone third-party interoperability testing, backward-compatibility testing, and is verified to have functional Wi-Fi features. The WFA certifications have

**Table 5.7**
DFS Requirements

| *Parameter* | *FCC* | *ETSI* |
|---|---|---|
| CAC Time | 60s | 10 minutes in 5600-5650 MHz and 10s elsewhere |
| Off-channel CAC time | Not allowed | 1 hour to 24 hours in 5600-5650 MHz and 6 minutes to 4 hours elsewhere |
| Radar detection threshold | -64 dBm for EIRP>= 23 dBm; -62 dBm for EIRP< 23 dBm and PSD< 10 dBm/MHz | -62+10-EIRP+G (dBm), where G is Rx antenna gain in dBi |
| Radar detection probability during CAC | 90% | 99.99% in 5600-5650 MHz and 60% elsewhere |
| Radar detection probability during in-service monitoring | 80% | 60% |
| Channel move time | 10s | 10s |
| Channel closing Tx time | First 200 ms for data traffic followed by 60 ms for control and management traffic | 1s |
| Nonoccupancy period | 30 minutes | 30 minutes |

today earned industry-wide recognition and awareness among consumers. The WFA has also largely succeeded in promoting the Wi-Fi brand over the years, while continuously improving the security aspects, breadth, and depth of interoperability testing. The WFA offers different Wi-Fi certifications for varied applications, they mainly focus on the functionality aspects with relevance to interoperability, and do not place much emphasis on performance verification. Below is a brief list of Wi-Fi certifications closely related to the topics covered in this book.

- *Wi-Fi Certified WPA3*: Brings the latest Wi-Fi security capabilities and enhances network security protection.
- *Wi-Fi Agile Multiband*: Enables STAs and APs to exchange information so the Wi-Fi network can guide client devices to connect to the best AP, frequency band and channel. This certification is largely based on features in the 802.11k, 802.11v, and 802.11r standards.
- *Wi-Fi Certified 6*: Provides higher capacity, performance, and improves efficiency based on the 802.11ax standard.

- *Wi-Fi Optimized Connectivity*: Enables faster network discovery and roaming using fast initial link setup (FILS), better network selection using link quality metric assessment, and reduces management frame overhead.
- *Wi-Fi Enhanced Open*: Adds encryption to open networks without the need for user intervention.
- *Wi-Fi Easy Connect*: Simplifies device provisioning and configuration using device provision protocol (DPP).

Table 5.8 captures the minimal set of mandatory features [8] for obtaining Wi-Fi 6 certification for both the AP and STA roles. Although several features described earlier in this book are optional for Wi-Fi 6 certification, note that it is common for WFA to begin a certification program for any new Wi-Fi generation with a small subset of features and then over time introduce new certification programs that expand the set of mandatory features.

**Table 5.8**

Wi-Fi 6 Mandatory Features

| *Feature* | *AP Role* | *STA Role* |
|---|---|---|
| 20 MHz bandwidth support in 2.4-GHz | Mandatory | Mandatory |
| 20, 40, 80 MHz bandwidth support in 5 GHz | Mandatory | Mandatory |
| Two spatial streams Tx and Rx | Mandatory | Mandatory |
| MCS 0 to 7 Tx and Rx | Mandatory | Mandatory |
| LDPC | Mandatory | Mandatory |
| SU beamformer | Mandatory if NSS>= 4 | Optional |
| SU beamformee | Optional | Mandatory |
| Downlink OFDMA | Mandatory | Mandatory |
| Uplink OFDMA | Mandatory | Mandatory |
| Downlink MU-MIMO | Mandatory | Mandatory |
| OMI | Mandatory | Mandatory |
| Individual TWT | Mandatory | Optional |
| WPA3 | Mandatory | Mandatory |
| Agile Multiband | Mandatory | Mandatory |
| BSS coloring | Mandatory | Mandatory |
| Spatial reuse | Optional | Optional |
| Duration-based RTS-CTS | Optional | Mandatory |
| MU EDCA | Optional | Mandatory |
| EMA with profile periodicity = 1 | Optional | Mandatory |

## 5.5 REFERENCES

[1] "IEEE Standard for Information Technology–Telecommunications and Information Exchange between Systems Local and Metropolitan Area Networks–Specific Requirements Part 11: Wireless LAN Medium Access Control (MAC) and Physical Layer (PHY) Specifications Amendment 1: Enhancements for High-Efficiency WLAN," *IEEE Std 802.11ax-2021 (Amendment to IEEE Std 802.11-2020)* (2021), pp. 1–767, DOI: 10.1109/IEEESTD.2021.9442429.

[2] "IEEE Standard for Information Technology–Telecommunications and Information Exchange between Systems - Local and Metropolitan Area Networks–Specific Requirements - Part 11: Wireless LAN Medium Access Control (MAC) and Physical Layer (PHY) Specifications," *IEEE Std 802.11-2020 (Revision of IEEE Std 802.11-2016)* (2021), pp. 1–4379, DOI: 10.1109/IEEESTD.2021.9363693.

[3] *Code of Federal Regulations, Title 47 CFR 15.247*, Federal Communications Commission, 2021.

[4] *ETSI EN 300 328 V2.2.2: Wideband transmission systems; Data transmission equipment operating in the 2.4 GHz band; Harmonised Standard for access to radio spectrum*, European Telecommunications Standards Institute, 2019.

[5] *Code of Federal Regulations, Title 47 CFR 15.407*, Federal Communications Commission, 2021.

[6] *ETSI EN 301 893 V2.1.1: 5 GHz RLAN; Harmonised Standard covering the essential requirements of article 3.2 of Directive 2014/53/EU*, European Telecommunications Standards Institute, 2017.

[7] *KDB 905462: Compliance measurement procedures for unlicensed-National information infrastructure devices operating in the 5250-5350 MHz and 5470-5725 MHz bands incorporating dynamic frequency selection*, Federal Communications Commission, Apr. 2016.

[8] Aruba Networks, *The Wi-Fi Market and the genesis of 802.11ax - Aruba Networks*, URL: https://www.arubanetworks.com/assets/wp/WP_802.11AX.pdf.

# Chapter 6

# Wi-Fi 6E

In the past 20 years, the demand for Wi-Fi has grown continuously with more than 15 billion active Wi-Fi enabled devices in the world today. With so many Wi-Fi deployments and devices, congestion is becoming an issue. While the 2.4-GHz band is already congested in many areas, the 5-GHz band is also expected to face similar congestion challenges within the next decade. Realizing the tremendous Wi-Fi growth and strong consumer demand for Wi-Fi, in April 2020, the Federal Communications Commission (FCC) voted to open up 1200 MHz of additional spectrum in the 6-GHz band (5.925-7.125 GHz) for unlicensed use. Quickly following this, the Wi-Fi industry members in the Institute of Electrical and Electronic Engineers (IEEE) and Wi-Fi Alliance (WFA) made use of this opportunity to unanimously agree upon adopting Wi-Fi 6 as the minimum supported Wi-Fi generation in the 6-GHz band, thereby eliminating overheads of backward compatibility. Few new features and changes specific to the 6-GHz band were added in the IEEE 802.11ax standard [1] to reduce management traffic, improve network discovery time, and provide superior user experience. In January 2021, WFA extended the Wi-Fi 6 certification program to the 6-GHz band along with certain new features under the moniker Wi-Fi 6E. Following the FCC in the United States, several countries such as Brazil, Canada, Chile, Costa Rica, European Union (EU), Guatemala, Honduras, Morocco, Norway, Peru, Saudi Arabia, South Korea, UAE, and UK have also opened up portions of the 6-GHz band for Wi-Fi. More countries are expected to follow suit in the coming years.

## 6.1 6-GHZ CHANNELIZATION

The 6-GHz band has a total of sixty nonoverlapping 20-MHz channels [2] and is divided into four Unlicensed National Information Infrastructure (U-NII) bands as follows:

1. U-NII-5 (5925 to 6425 MHz) comprising twenty-five 20-MHz channels.
2. U-NII-6 (6425 to 6525 MHz) comprising five 20-MHz channels.
3. U-NII-7 (6525 to 6875 MHz) comprising seventeen and a half 20-MHz channels.
4. U-NII-8 (6875 to 7125 MHz) comprising twelve and a half 20-MHz channels.

The channel numbering in 6-GHz ranges from 1 to 233 and the channel center frequency can be derived using (6.1).

$$\begin{aligned}\text{Channel center frequency (MHz)} = {} & \text{Channel starting frequency (MHz)} + \\ & 5 \cdot \text{Channel number}\end{aligned} \tag{6.1}$$

where the channel starting frequency is defined by the operating class. Table 6.1 details the channel starting frequency and allowed channel numbers for each operating class and bandwidth. The operating class 136, which enables use of the frequency spectrum 5925 to 5945 MHz, is not supported by many implementations as it is instead used as a guard band to minimize emissions outside the 6-GHz band. Note that some channels span across two U-NII bands. For example, channel 185 in operating class 131 spans across U-NII-7 and U-NII-8 bands. For such channels, the tighter of the regulatory rules of the two bands apply.

## 6.2 AP DISCOVERY

The AP discovery process in 5-GHz band if extended to the 6-GHz band would result in high scan times for STAs owing to the large number of 6-GHz channels. Also, active scanning process results in significant amount of management traffic. Given that 6 GHz is a new spectrum, Wi-Fi 6E introduces new AP discovery mechanisms to minimize scan time and to reduce management traffic. An AP capable of 6-GHz operation is classified into one of the below two types and the applicable AP discovery mechanism depends on the AP type.

**Table 6.1**
Operating Classes and Channel Numbers

| *Operating Class* | *Bandwidth (MHz)* | *Channel Starting Frequency (MHz)* | *Channel Numbers* | *Number of Channels* |
|---|---|---|---|---|
| 131 | 20 | 5950 | 1, 5, 9, ..., 229, 233 | 59 |
| 132 | 40 | 5950 | 3, 11, 19, ..., 219, 227 | 29 |
| 133 | 80 | 5950 | 7, 23, 39, ..., 199, 215 | 14 |
| 134 | 160 | 5950 | 15, 47, 79, 111, 143, 175, 207 | 7 |
| 136 | 20 | 5925 | 2 | 1 |

1. 6-GHz AP: A 6-GHz capable AP with colocated 2.4-GHz and/or 5 GHz radio.
2. 6-GHz only AP: An AP that is capable of operating only in the 6-GHz band.

### 6.2.1 Minimizing Probe Traffic

To minimize management traffic related to AP discovery, Wi-Fi 6E prohibits a STA from transmitting a broadcast probe request with wildcard service set identifier (SSID) in the 6-GHz band. Active scan in the 6-GHz band is permitted using unicast probe requests or broadcast probe requests with a specified SSID field. However, a STA is aware of the SSID or BSS identifier (BSSID) only in the case of a known Wi-Fi network. Discovery of unknown Wi-Fi networks in 6 GHz relies on an out-of-band discovery mechanism for 6-GHz APs and a fast passive scanning mechanism for 6-GHz only APs.

#### 6.2.1.1 Short SSID

When a STA scans for networks with known SSID, it specifies the SSID in the probe request. The SSID field size is variable and can be as long as 32 bytes. Wi-Fi 6E introduces an optional short SSID feature, that enables a STA to specify a new 4-byte field called short SSID instead of the full SSID in its probe request. The short SSID is simply the 4-byte hash sum of the actual SSID. The AP upon receiving a probe request with the short SSID element responds with a probe response if the short SSID value in the probe request matches with any of the corresponding short SSIDs it is servicing.

### 6.2.2 Out-of-Band Discovery of a 6-GHz AP

Wi-Fi 6E requires a 6-GHz AP to advertise all its 6-GHz BSSs in the colocated 2.4-GHz or 5 GHz basic service set (BSS) using the reduced neighbor report (RNR) IE. The RNR IE is present in 2.4-GHz and 5-GHz beacons and probe responses, and it contains one or more neighbor AP information fields, that specify the BSS information related to a group of BSSs having the same channel and operating class. Figure 6.1 shows the format of the RNR IE. Each neighbor AP information field specifies the 6-GHz primary 20-MHz channel number, operating class, and a TBTT information field for each 6-GHz BSS belonging to the group. The TBTT information field comprises of a 48-bit BSSID, a 32-bit short SSID, an 8-bit 20 MHz PSD subfield, and an 8-bit BSS parameters subfield. The BSS parameters subfield indicates information such as whether the BSS is part of a multiple BSSID set, whether the BSS is a transmitted BSSID or a nontransmitted BSSID, and whether the SSID is same as the colocated 2.4-GHz or 5-GHz BSS. The 20 MHz PSD subfield indicates the AP's power spectral density (PSD) limit in the primary 20-MHz 6-GHz channel.

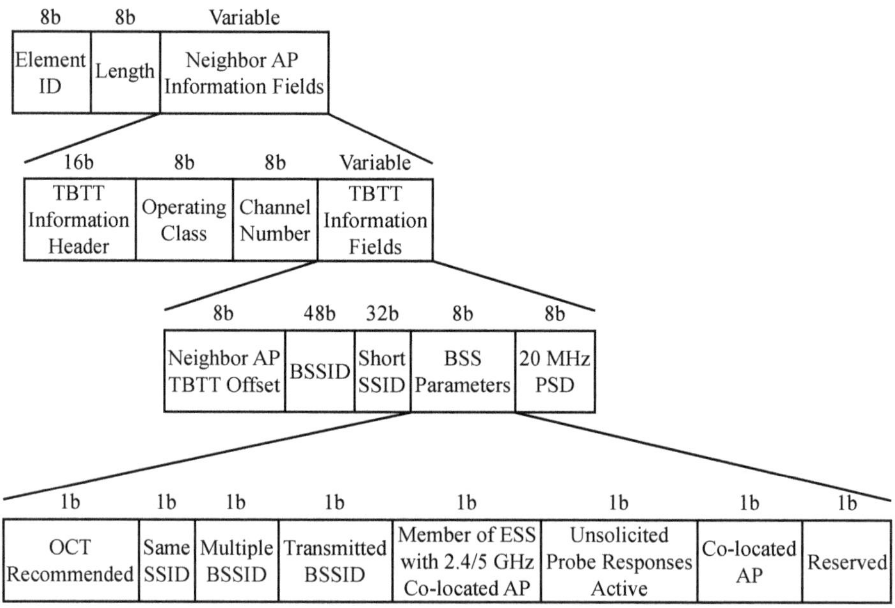


**Figure 6.1** Reduced neighbor report IE format.

After AP discovery, if the STA decides to connect to the 6-GHz BSS, then it goes to the primary 20-MHz channel of the 6-GHz BSS and sends a unicast probe request with the specified SSID field. From the out-of-band discovery process, the STA is aware of the SSID, BSSID, primary 20-MHz channel number and the PSD limit in 6-GHz. The rest of the connection process is similar to Wi-Fi 6. Out-of-band discovery enables a STA to discover a 6-GHz AP in the same time it takes to discover a 5-GHz AP. Moreover, out-of-band discovery pushes all the discovery related management traffic to 2.4-GHz and 5-GHz band, thereby keeping the 6-GHz band pristine for the purpose of data traffic.

### 6.2.3 Fast Passive Scanning of a 6-GHz Only AP

Discovery of a 6-GHz only AP relies on passive scanning of the STA in the 6-GHz band. To enable faster passive scanning by STAs, 6-GHz only APs are required to choose their primary 20-MHz channel from a smaller subset of channels called preferred scanning channels (PSCs). There are a total of 15 PSCs defined with a channel spacing of 80 MHz and their channel numbers are $5, 21, 37, 53, \ldots, 229$.

A 6-GHz only AP is also required to transmit either an unsolicited broadcast probe response frame or a fast initial link setup (FILS) discovery frame every 20 time units (TUs) or lower. FILS Discovery frames are like small beacon frames that include minimal information for AP discovery such as SSID, BSSID, RNR IE, operating channel width, and number of spatial streams. A STA can therefore discover a 6-GHz only AP on a channel by passive scanning for 20.48 ms. For example, if the STA dwell time is 40 ms and STA channel switch time is 10 ms, then fast passive scanning of all PSCs takes $(40 + 10) \cdot 15\text{ ms} = 750\text{ ms}$. A 6-GHz only AP is also permitted to send a broadcast probe response frame in response to a unicast probe request frame with specified SSID. This broadcast probe response frame can also be counted towards meeting the requirement of an unsolicited broadcast probe response frame every 20 TUs. This elegant technique helps lower the amount of probe traffic as other listening STAs on the same channel receive a probe response without having to send a probe request.

## 6.3 EMA SUPPORT IN 6 GHZ

Support for enhanced multi-BSSID advertisement (EMA) with a profile periodicity of 1 is mandatory for both APs and STAs in the 6-GHz band. This feature is useful when there are multiple virtual APs (VAPs) in the 6-GHz band since beacons and probe responses are limited to only the transmitted BSSID, thereby reducing management traffic airtime.

## 6.4 SECURITY MODES IN 6 GHZ

WPA3-SAE, WPA3-Enterprise and OWE are the only supported security modes in 6-GHz band. OWE provides an encrypted wireless link even for open authentication networks. Since there are no legacy STAs in the 6-GHz band, OWE transition mode and WPA3 transition mode are not needed.

## 6.5 BEACON ADVERTISEMENT AND PHY RATES IN 6 GHZ

Since the 6-GHz band does not support previous Wi-Fi generations, the beacons and probe responses in 6 GHz do not contain HT capabilities IE, HT operational IE, VHT capabilities IE and VHT operational IE. A new HE 6 GHz band capabilities IE is added instead in the 6-GHz beacons and probe responses to advertise some of the medium access control (MAC) parameters. Figure 6.2 shows the various fields in HE 6 GHz band capabilities IE.

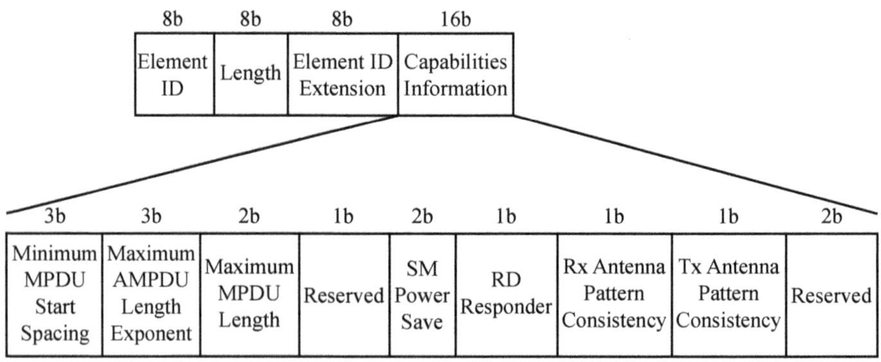


**Figure 6.2** HE 6 GHz band capabilities IE format.

Moreover, the 6-GHz operation information field in HE operation IE specifies the AP's primary channel, channel center frequency, bandwidth, and minimum physical (PHY) rate to be used by STAs. Figure 6.3 shows the different subfields of 6 GHz operation information field. The primary channel subfield indicates the channel number of the primary channel and the channel center frequency segment 0 subfield specifies the channel center frequency index for the 20-MHz, 40-MHz, 80-MHz or 160-MHz channel on which the BSS operates. The channel width subfield of the control subfield specifies the bandwidth of the channel and is set to 0 for 20 MHz, 1 for 40 MHz, 2 for 80 MHz, and 3 for 160 MHz. The minimum rate subfield indicates the minimum PHY rate in units of 1 Mbps to be used by STAs for sending PPDUs and the chosen minimum PHY rate should have MCS $< 3$ and NSS $< 3$.

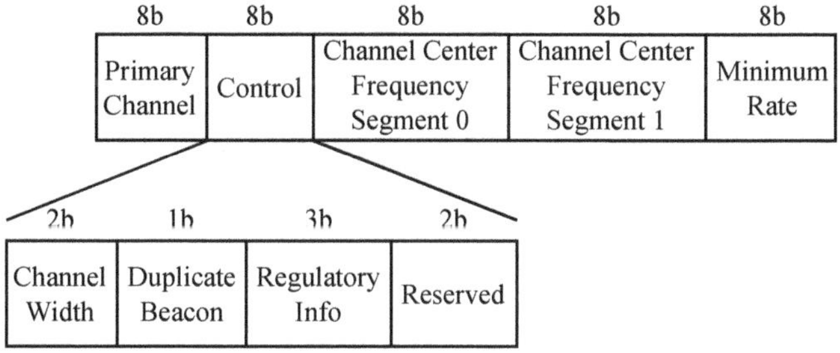


**Figure 6.3** 6 GHz operation information field format.

Although both AP and STA operating in 6 GHz are expected to use HE PHY rates for all transmissions, the standard allows an exception to use legacy PPDU and non-HT legacy duplicate PPDU. This exception is intended to be used mainly for control and management frames such as ACK, BA, RTS, CTS and beacon frames.

## 6.6 ON CHANNEL TUNNELING

On channel tunneling (OCT) is an optional feature that allows a multiband capable STA and a multiband capable AP to exchange management frames (also called MMPDUs) intended for one frequency band in a different frequency band. In the context of a 6-GHz AP, a triband (2.4/5/6 GHz) capable STA can encapsulate a MMPDU meant for 6-GHz BSS in a special action frame called OCT MMPDU and send it to a triband capable AP in the 2.4- or 5-GHz band. The 2.4-GHz or 5-GHz BSS upon receiving the OCT MMPDU will forward it to the 6-GHz BSS for appropriate action. Similarly a triband capable AP can encapsulate a MMPDU originating from 6-GHz BSS in an OCT MMPDU and send it in the 2.4 or 5-GHz band to a triband capable STA. Essentially, OCT feature enables out-of-band authentication and association of a Wi-Fi 6E STA with a 6-GHz AP and this is illustrated in Figure 6.4. After completing association in the 2.4- or 5-GHz band, the STA can switch to the 6-GHz BSS channel and directly start sending data traffic in 6 GHz.

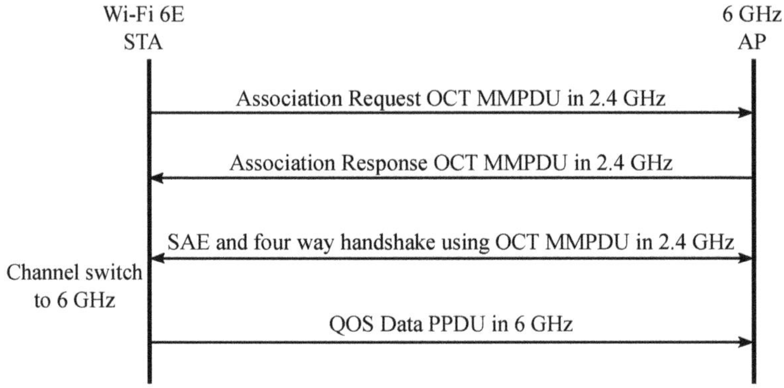


**Figure 6.4** Example of OCT operation.

Support for OCT feature is indicated by a STA in the OCT not supported subfield of multi-band control field in multi-band IE. An AP indicates OCT feature support in the OCT field of extended capabilities IE.

## 6.7 FAST INITIAL LINK SETUP

Fast initial link setup (FILS) is a 802.11ai feature that is part of the Wi-Fi optimized connectivity experience (OCE) certification program. FILS reduces the protocol overhead of link setup and makes the connection setup process faster. The goal of FILS is to enable a client to complete the link setup and obtain a valid IP address within 100 ms. FILS is especially useful in high-density enterprise environments with a large number of mobile users in the network, wherein a significant fraction of airtime is spent on network discovery and authentication.

FILS makes the network discovery process efficient by requiring APs to send broadcast probe response frame in response to a unicast probe request frame. Also, APs are encouraged to periodically send broadcast probe response or FILS discovery frames so clients can passively discover APs around them. Clients are also required to list the APs they have already discovered in their active scan probe request frame so that those APs do not respond, thus saving on airtime. Some of these ideas on improving the AP discovery process are adopted in Wi-Fi 6E fast passive scanning.

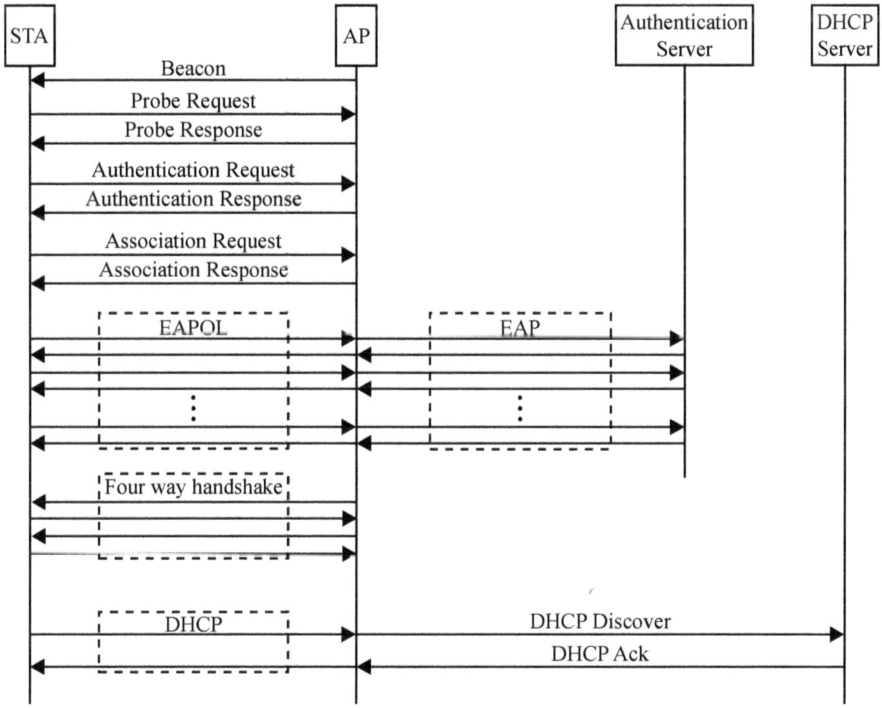


**Figure 6.5** Steps in 802.1X connection establishment and IP address assignment.

After network discovery, a client has to complete 802.1X authentication and obtain an IP address from a dynamic host configuration protocol (DHCP) server. As shown in Figure 6.5, there are more than 25 frames from a probe response to obtaining an IP address. With FILS, a client still needs to complete the full 802.1X EAP exchange for the first time it connects to a network, but the authentication server remembers this information for several days. For subsequent connections to the network, the FILS authentication method based on the EAP reauthentication protocol revives the full authentication using just two frame exchanges. Finally, the FILS association request and FILS association response frames enable the AP to request the DHCP server for an IP address on behalf of the client and issue the granted IP address back to the client. So, the IP address assignment happens concurrently with client association. Figure 6.6 shows the various steps in the FILS process.

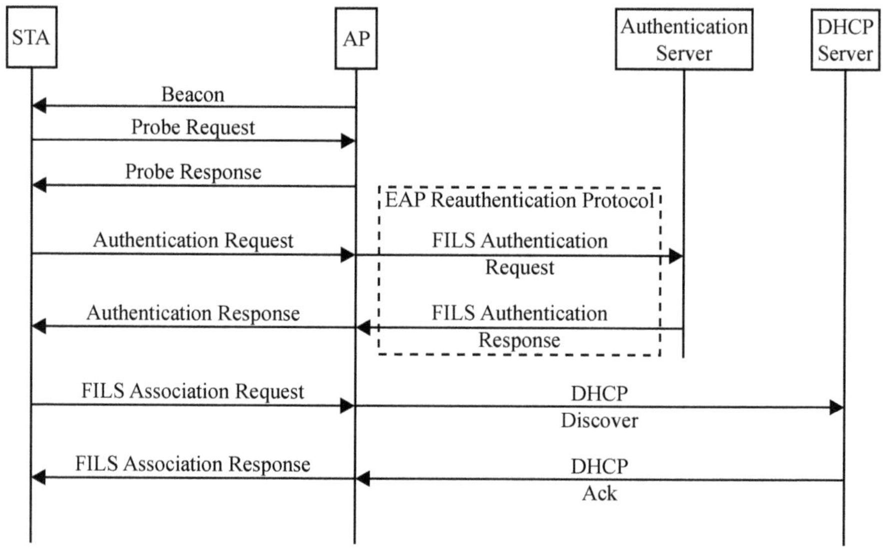


**Figure 6.6** Steps in FILS with concurrent IP address assignment. (*Source:* Nanocell Networks Pvt. Ltd.)

### 6.7.1 FILS and Fast BSS Transition

Although many of the objectives of FILS are similar to 802.11r fast BSS transition (FT) roaming, there are some important differences to note. FT roaming concerns fast roaming from one AP to another AP of the same network by skipping 802.1X authentication and four-way handshake, but it is useful only as long as the client is continuously connected to the same network. For example, if a user connects to a network in the morning, leaves the network at noon, and then arrives later in the afternoon to connect to the network, FT roaming does not help in this scenario but FILS will be helpful. FILS is therefore relevant when there are many mobile users frequently entering and leaving the network. Moreover, FT roaming does not help with expediting IP address assignment. Although FILS support is not required as part of Wi-Fi 6E certification, the absence of legacy devices in the 6-GHz band presents an opportunity to take advantage of the many benefits of the FILS feature.

## 6.8 6-GHZ REGULATORY REQUIREMENTS

The list of available channels and maximum Tx power limits for indoor and outdoor operation in 6 GHz vary by country. Many countries are in the process of evaluating whether to open up 6-GHz band for Wi-Fi. Although some countries have decided to open up the 6-GHz band for Wi-Fi, the exact regulatory requirements are still evolving at the time of writing this book. In some countries, 5G cellular technology is also competing for the U-NII-6, U-NII-7, and U-NII-8 spectrum. So, it is possible that U-NII-5 might be the globally available band. The high-level regulatory requirements for FCC and ETSI are described in this section, but some of these could change over time.

### 6.8.1 FCC

The 6-GHz spectrum in the United States is primarily allocated to point-to-point fixed service (FS) links and fixed-satellite service (FSS) earth to space uplink. There are thousands of these incumbents already deployed in the United States. The U-NII-5 and U-NII-7 bands are used by FSS links for communications to geostationary satellites, point-to-point microwave links for wireless backhaul and utilities, public safety, and emergency services. The U-NII-6 and U-NII-8 bands are used for mobile services, for example trucks that relay signals back to a TV studio. To mitigate the interference to satellite receivers of FSS incumbents, all outdoor APs are required to limit their maximum EIRP at any elevation angle above 30 degrees from the horizon to 21 dBm. To protect the FS incumbents, FCC has defined two classes of 6-GHz operation for APs called low power indoor (LPI) operation and standard power (SP) operation.

#### 6.8.1.1 Low Power Indoor Operation

The LPI class of APs can be used only indoors. LPI APs are free to operate across the entire 6-GHz band with a maximum equivalent isotropically radiated power (EIRP) limit of 30 dBm and a PSD limit of 5 dBm/MHz. The low PSD limit of 5 dBm/MHz is the likely limiting factor for Tx power in most indoor AP designs. This is illustrated in Table 6.2, which shows the LPI AP Tx power limit calculation for different bandwidths assuming a directional gain of 5 dB (for example $2 \times 2$ AP with 2 dBi antenna gain). It is evident from Table 6.2 that AP operation in higher bandwidth is clearly preferred owing to the higher Tx power limit.

Clients are not allowed to transmit anything on a 6-GHz channel without first confirming the presence of an AP operating on that channel by either passive scanning or using out-of-band discovery. Clients connected to an LPI AP have a 6 dB lower Tx power limit (maximum EIRP of 24 dBm and PSD limit of -1 dBm/MHz) than the AP.

**Table 6.2**

Transmit Power Limit for LPI Class AP with 5-dB Directional Gain

| *Bandwidth (MHz)* | *Transmit Power Limit (dBm)* |
|---|---|
| 20 | $5 + 10 \cdot \log_{10}(20) = 18$ |
| 40 | $5 + 10 \cdot \log_{10}(40) = 21$ |
| 80 | $5 + 10 \cdot \log_{10}(80) = 24$ |
| 160 | $30 - 5 = 25$ |

In order to simplify the enforcement of indoor operation, LPI class APs cannot have weather resistant enclosures, cannot be battery-powered and must use integrated antennas with no provision for external antennas.

#### 6.8.1.2 Standard Power Operation

SP class of operation enables both indoor and outdoor APs to operate under the control of an automatic frequency coordination (AFC) system in the U-NII-5 and U-NII-7 bands at a maximum possible EIRP limit of 36 dBm and a PSD limit of 23 dBm/MHz. Figure 6.7 shows the various components of an AFC system. Before beginning SP operation, AFC requires an SP AP to connect to an AFC operator and provide its geolocation and location uncertainty measurement with 95% confidence. SP APs have the option to either integrate geolocation functionality or have a secure physical link to an external geolocation device. In the case of external geolocation device, the distance from the AP to the external geolocation device must be included in the location uncertainty measurement. The geolocation measurement can use any technology such as global positioning system (GPS), but it has to be an autonomous measurement and cannot be manually entered during deployment to avoid the possibility of human error. The AFC operator maintains a database of the licensed FS operators including each of their physical location, Tx power, bandwidth, antenna height, and so on. This information is updated every 24 hours from the FCC universal licensing system (ULS) database. The AFC operator upon receiving the AP's geolocation and location uncertainty responds to the AP with a list of allowed channels and maximum allowed EIRP in those channels based on the list of FS operators near the AP's location. The AP can then begin operation in those channels subject to the authorized maximum EIRP limit.

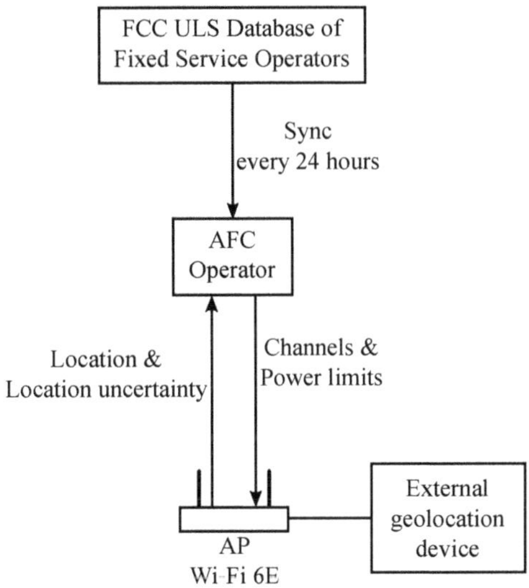


**Figure 6.7** AFC system components.

The WFA has developed a detailed specification [3] for the interface between AP and AFC operator. The FCC has provided some guidance [4] on the acceptable interference level at the FS incumbent and detailed how to perform the propagation loss calculations, but the exact details are still being formulated by a multistakeholder group. The AFC operator can be physically anywhere in the United States and is subject to oversight by the FCC. Clients connected to an SP AP have to use a 6 dB lower Tx power than their AP's authorized Tx power limit as provided by the AFC operator.

#### 6.8.1.3 Out-of-Band Emission Limit

FCC specifies an out-of-band emission limit of -27 dBm/MHz for emissions outside the 5.925-7.125 GHz band, which is applicable to both LPI and SP class of operation.

### 6.8.2 ETSI

ETSI [2, 5] allows operation only in the 5.945-6.425 GHz band. This 480 MHz of spectrum offers twenty-four 20-MHz channels, twelve 40-MHz channels, six 80-MHz channels and three 160-MHz channels. ETSI has defined two classes of 6-GHz device operation called very low power (VLP) and LPI.

#### 6.8.2.1 Very Low Power

The VLP class is applicable to both indoor and outdoor operation. It allows a maximum EIRP limit of 14 dBm, and a PSD limit of 1 dBm/MHz while requiring an out-of-band emission limit of -45 dBm/MHz for emissions below 5935 MHz.

#### 6.8.2.2 Low Power Indoor

As in FCC, the LPI class is applicable for indoor operation only. LPI class permits a maximum EIRP limit of 23 dBm, a PSD limit of 10 dBm/MHz, and requires an out-of-band emission limit of -22 dBm/MHz for emissions below 5935 MHz.

## 6.9 WI-FI 6E CERTIFICATION

Wi-Fi 6E is a certification option under the Wi-Fi 6 certification program that extends Wi-Fi 6 certification to the 6-GHz band, deprecates older Wi-Fi generation support in 6 GHz, and adds new features to keep network operation in 6 GHz highly efficient. Table 6.3 lists the minimum set of mandatory features to obtain Wi-Fi 6E certification.

## 6.10 WI-FI 6E IMPLEMENTATION CHALLENGES

Addition of 6-GHz band support presents few hardware design challenges. One challenge is antenna design as it is difficult to design a single, wide-band antenna covering both 5-GHz and 6-GHz bands without making some tradeoffs on technical specifications. If separate antennas are designed for 5 GHz and 6 GHz, space constraints may become a challenge especially for client devices. The other aspect is deciding the number of radios needed. For an AP, it would be more appropriate to have three independent radios supporting triband concurrent operation since the

**Table 6.3**
Wi-Fi 6E Mandatory Features

| *Feature* | *AP Role* | *STA Role* |
|---|---|---|
| Wi-Fi 6 certification in all supported bands | Mandatory | Mandatory |
| Out-of-band discovery | Mandatory for 6-GHz AP | Mandatory if multiple bands are supported |
| Fast passive scanning | Mandatory for 6-GHz only AP | Mandatory |
| EMA with profile periodicity = 1 | Mandatory | Mandatory |
| OWE | Optional | Optional |
| FILS | Optional | Optional |
| OCT | Optional | Optional |

majority of clients do not support the 6-GHz band. However, three radios increase both system cost and system power consumption. This makes it difficult to fit within either the 802.3af or 802.3at power over Ethernet (PoE) power budgets, which are more common among network switches.

## 6.11 160-MHZ MULTIPLE AP DEPLOYMENT IN 6 GHZ

Although the 160-MHz bandwidth mode was introduced in Wi-Fi 5, it is rarely used in practice owing to the limited 160 MHz-channel availability in 5-GHz band and DFS complexity. As per the FCC, there are only two 160-MHz channels in the 5-GHz band. This restricts 160-MHz usage in 5-GHz only to homes where there are only one or two APs. Clearly 160 MHz in the 5-GHz band is not going to be practical in enterprise deployments or multidwelling units (MDUs). Now with seven 160-MHz channels available in the 6-GHz band with LPI class of operation in the United States, is it possible to have an indoor 160 MHz multiple AP enterprise deployment?

To answer this question, it will be useful to borrow some concepts from cellular networks. Let us first define a cell to be the coverage area of an AP wherein a client can establish connection and obtain a minimum throughput. This minimum throughput depends on the use case. The coverage area largely depends on the minimum of the Tx power of both the AP and client. Owing to the omnidirectional nature of RF propagation, the shape of the cell is a circle assuming no physical obstructions like walls. To fill the entire area of an enterprise venue, multiple APs or cells are required and to provide the best performance it is necessary to maximize the number of cells that can be packed in a venue. The best cell shape that can pack a given area with the highest cell packing density is known to be the hexagon shape. Therefore, it is common practice to approximate a cell with a hexagon shape that is inscribed inside the circular-shaped cell. When two hexagons are placed next to each other with their sides touching, this creates a natural overlap of the actual circular cell area. Cell overlap is necessary to facilitate seamless roaming of a client.

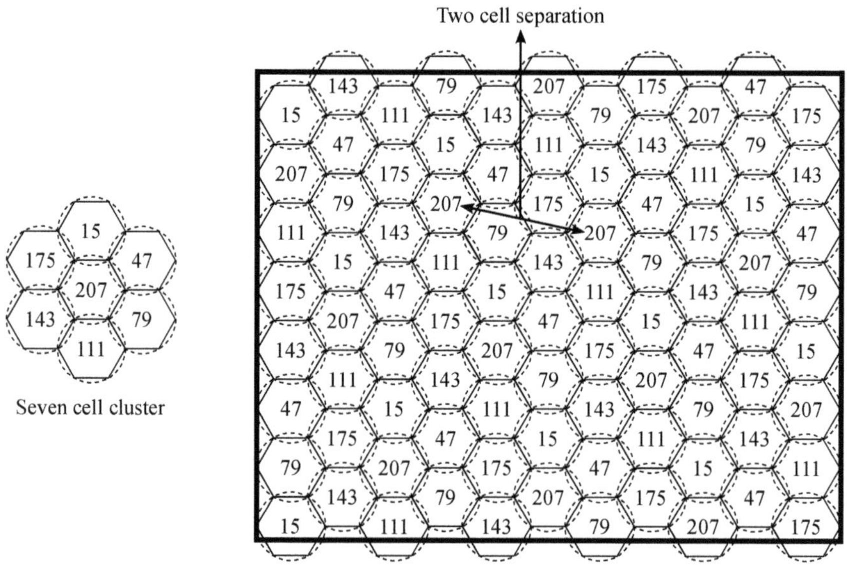


**Figure 6.8** Example of 160-MHz enterprise deployment in 6 GHz.

To mitigate interference between APs in a venue, the cell overlaps across the entire venue should involve only nonoverlapping channels. Since there are six sides to a hexagon, it can be surrounded with six hexagonal cells, one on each side. Let us define this set of one hexagonal cell surrounded by six hexagonal cells as a cluster. The seven 160-MHz channels (15, 47, 79, 111, 143, 175, 207) available in FCC LPI class can be assigned to each of the seven cells in the cluster and this cluster can be repeated multiple times to cover the entire area of the venue as shown in Figure 6.8. The channel assignment shown in Figure 6.8 demonstrates a minimum of two cell separation between any two cells with the same channel. With proper channel planning, it is therefore possible to have a 160-MHz multiple AP deployment in 6 GHz.

## 6.12 REFERENCES

[1] "IEEE Standard for Information Technology–Telecommunications and Information Exchange between Systems Local and Metropolitan Area Networks–Specific Requirements Part 11: Wireless LAN Medium Access Control (MAC) and Physical Layer (PHY) Specifications Amendment 1: Enhancements for High-Efficiency WLAN," *IEEE Std 802.11ax-2021 (Amendment to IEEE Std 802.11-2020)* (2021), pp. 1–767, DOI: 10.1109/IEEESTD.2021.9442429.

[2] *Wi-Fi 6E and 6 GHz Update, Version 1.0*, Wi-Fi Alliance, Mar. 2021.

[3] *AFC System to AFC Device Interface Specification*, Wi-Fi Alliance, Feb. 2021.

[4] *ET Docket No. 18-295: Unlicensed Use of the 6 GHz band*, Federal Communications Commission, Apr. 2020.

[5] *Draft ETSI EN EN 303 687: 6 GHz RLAN Harmonised Standard for access to radio spectrum*, European Telecommunications Standards Institute, 2021.

# Chapter 7

# Wi-Fi Deployment

This chapter is intended to provide an overview of the common challenges and various aspects involved in deploying Wi-Fi. The nature of challenges depends on the venue and service requirements. While each deployment venue could have certain unique challenges, there are broadly three categories of deployment venues as listed below:

1. Residential;
2. Enterprise indoor;
3. Outdoor.

The number of APs in a deployment, type of backhaul link, number of clients to be served, and the per-client performance requirement largely depend on the venue category. Sections below cover the important aspects for each venue category and also provide some guidance to help simplify the deployment.

## 7.1 RESIDENTIAL WI-FI DEPLOYMENT

The internet service to a residence is typically done over cable, digital subscriber line (DSL), or fiber, and the internet service provider (ISP) enforces a maximum download and upload speed depending on the internet service package. In most homes, the maximum download and upload speeds are not limited by Wi-Fi, but rather limited by the ISP. The number of APs in a residential Wi-Fi deployment typically ranges from 1 to 3 and is largely determined by coverage requirements.

Large homes, multistory homes, and homes with brick walls require more than one AP to provide coverage across the entire home. Using a single AP with higher transmit (Tx) power doesn't always help increase the coverage range of a AP because the client's Tx power could become the bottleneck. Therefore multiple APs often provide a better coverage compared to a single AP with high Tx power. In a multiple AP residential deployment, one of the APs is connected to the wired backhaul link, that provides internet service to the home and this AP is called a root AP. The remaining APs typically use Wi-Fi as the backhaul link connecting to the root AP or use Ethernet to connect to the root AP. Since older homes lack Ethernet cabling across the home, consumers living in such homes prefer wireless backhaul to avoid the effort of installing Ethernet cabling. The APs that use Wi-Fi backhaul and also provide wireless service are called mesh APs or repeater APs. Mesh APs can either connect directly to the root AP or connect to another intermediate mesh AP.

### 7.1.1 Mesh

Mesh essentially forms a multiple hop Wi-Fi link to extend the coverage range and avoids the hassle of cable wiring providing more flexibility for AP placement. A mesh AP simultaneously acts as a client as well as an AP and repeats frames received in one link over the other link as illustrated in Figure 7.1. There are multiple protocols available for mesh. Today, there are mesh products broadly based on two types of mesh protocols as listed below.

1. *802.11s mesh*: Based on the IEEE 802.11s amendment and mesh BSS network topology. It supports multiple wireless backhaul paths for mesh AP making it fault-tolerant in the event of a mesh AP failure.

2. *Wireless distribution system (WDS) mesh*: Based on the WDS network topology and uses a simpler four address frame format. In the event of a mesh AP failure, there will be service disruption until an alternate connection is established.

For small networks such as in a residential deployment, the complexity and sophistication of 802.11s mesh is not necessary and the WDS style of mesh can largely meet the needs of residential use cases. Since every device vendor uses a proprietary style of mesh, interoperability is an issue between mesh APs from different vendors. In 2018, WFA introduced the EasyMesh certification program based on the WDS style of mesh to ensure mesh interoperability.

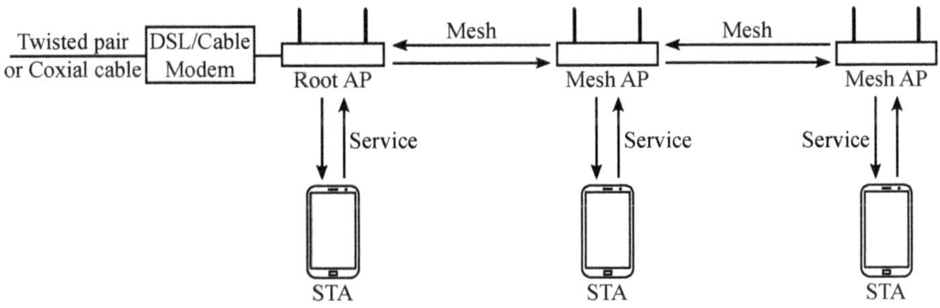


**Figure 7.1** Mesh Wi-Fi.

Most mesh AP implementations use the same radio for backhaul and service (or fronthaul) links, resulting in time multiplexing between the two links. Hence, the capacity of such mesh AP implementations decreases linearly with the number of hops required to reach the STA from root AP. Similarly, the packet latency as measured from root AP to STA increases with the number of hops. Some mesh implementations avoid this penalty by using a different radio for backhaul and fronthaul links, and assign different channels, but this comes with an increased cost. With 6 GHz spectrum available for Wi-Fi in some countries, one possibility in a triband Wi-Fi 6E AP is to dedicate the 6-GHz radio for mesh backhaul while the 2.4-GHz and 5-GHz radios are dedicated for wireless service.

### 7.1.2 AP Placement

APs should be placed closer to where users or client devices are mostly located in the home to get the best performance. APs should not be placed close to metal objects or near non-Wi-Fi interference sources like a microwave, cordless phone or Bluetooth speaker. Often, the AP placement in a home is constrained by the location of electrical outlets.

### 7.1.3 Channel Bandwidth Selection

It is often good practice to scan the environment for Wi-Fi APs nearby in all channels as there could be Wi-Fi interference from neighboring homes. Since there are a few APs in any residence, it is preferable to use the widest possible bandwidth in all supported bands unless several channels are occupied by neighbor home APs.

## 7.2 ENTERPRISE INDOOR WI-FI DEPLOYMENT

Enterprise deployments are managed deployments either by an on-premise controller or a cloud based controller. The controller is mainly responsible for managing the APs in terms of configuring them and providing a single pane view of what is happening in the entire network. This includes visibility into the traffic statistics, the user experience per-client, the channels and airtime used by each AP, the areas of the network facing medium congestion and general health metrics of APs, such as power status, memory, CPU utilization, system crashes, and reboots. In addition, depending on the implementation, the controller may perform other functionality related to security, roaming, dynamic host configuration protocol (DHCP) server, radio resource management, and so on. A block diagram of the main components in an enterprise network are shown in Figure 7.2. All APs are connected using Ethernet cables to a network switch that provides both the wired backhaul link as well as power to the APs using power over Ethernet (PoE).

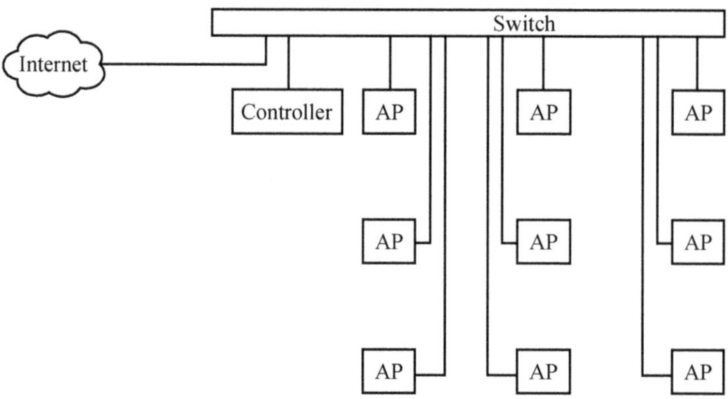


**Figure 7.2** Components of enterprise network.

### 7.2.1 Power over Ethernet

PoE is a standard that enables a single Ethernet cable to carry both data and electrical power to the AP. The device that provides power on the Ethernet cable is called power sourcing equipment (PSE) and the device that is powered by PoE is called powered device (PD). In the present context, the network switch acts as PSE and the APs are PDs. The IEEE has so far defined three standards: 802.3af,

802.3at, and 802.3bt, for delivering different power levels over Ethernet. Table 7.1 summarizes the important PoE parameters for different power types defined in the IEEE standards. Note that the maximum power available at the PD is lower than that delivered by the PSE due to the worse-case assumption of cable loss for a maximum supported cable length of 100 meters.

**Table 7.1**
PoE Power Types

| *Parameter* | *Type 1 (802.3af)* | *Type 2 (802.3at)* | *Type 3 (802.3bt)* | *Type 4 (802.3bt)* |
|---|---|---|---|---|
| Maximum power available at PD | 12.95 W | 25.5 W | 51 W | 71 W |
| Maximum power delivered by PSE | 15.4 W | 30 W | 60 W | 90 W |
| Power class levels | 1 to 3 | 1 to 4 | 1 to 6 | 1 to 8 |
| Supported cabling | Category 3 | Category 5 | Category 5 | Category 5 |

The actual power delivered by the PSE is negotiated between the PSE and PD using two handshakes. During the first handshake, the PSE sends a short time duration, low-voltage pulse and measures the current drawn during the pulse. The low voltage of this pulse is unlikely to damage a non-PoE device. The PD communicates a desired power level by presenting an appropriate resistance to the PSE. The PSE classifies the PD into a power class level depending on the current drawn but then starts delivering power to the PD limiting it to 802.3af Type 1. A second handshake is performed before the PSE classifies the PD to a higher power type. The second handshake can either repeat the same hardware pulse-based classification method or use a software-based link layer discovery protocol (LLDP). The LLDP involves certain frame exchanges over the Ethernet link to negotiate the desired power in 0.1 W steps while the hardware pulse based classification is more granular as defined by power class levels. Note that LLDP must be supported on both the PSE and PD for the second handshake to utilize the LLDP method.

### 7.2.2 Choosing AP Specification

Before getting to deployment, the technical specifications for the AP have to be decided first. This concerns the choice of Wi-Fi generation, number of frequency bands supported, number of Tx and receive (Rx) antennas, and so on. This section offers guidelines that can help in determining the AP specifications for a deployment.

It is recommended to choose the latest Wi-Fi generation for the AP as it will future-proof the network. As well, the latest generation hardware is expected to be faster owing to a faster process node for the application specific integrated circuit (ASIC). In certain deployments where hardware and software maturity is required on day 1 of deployment, an older yet mature AP model is typically preferred. The number of Tx and Rx antennas on the AP should be higher than the client's capabilities in order to experience multiuser MIMO (MU-MIMO) and transmit beamforming (TxBf) benefits. For optimal MU-MIMO benefits with $2 \times 2$ STAs, eight antennas are recommended at the AP with four antennas being the minimum requirement. The AP should at least have the same number of Tx and Rx antennas as the most dominant client type in the network so that the performance is not limited by the AP's capability. Several recently launched client devices support $2 \times 2$ spatial streams as a minimum. Given the low amount of spectrum in the 2.4-GHz band, it is desirable for the AP to support concurrent operation in both the 2.4-GHz and 5-GHz bands unless the deployment is expected to serve only 2.4-GHz capable clients. In the future, triband APs might become common as more Wi-Fi 6E capable client devices are introduced in the market. The AP should also be capable of handling the peak associated client count expected on a single AP in the network. Accordingly, the AP should have adequate memory to handle the peak client load. The maximum packets per second that can be processed by the AP should be higher than the expected peak load of small packets per second. Finally, the PoE power type required for optimum AP performance should be supported by the network switch and if not, the network switch should be upgraded.

### 7.2.3 Channel Bandwidth Selection

After finalizing the AP specifications, the channel bandwidth configuration for the AP has to be tentatively determined for better estimation of the total airtime consumption, which impacts the estimate for the number of APs. This tentative decision on channel bandwidth can be revisited after a detailed site survey, coverage and capacity planning exercise. While it is desired to configure all APs to the maximum supported channel bandwidth of the AP, the number of nonoverlapping channels reduces with increasing channel bandwidth. In the 5-GHz band, more than 60% of the spectrum is shared with radar and dynamic frequency selection (DFS) is required to operate in U-NII-2A and U-NII-2C bands. Some deployments prefer to disable DFS to avoid service disruption due to channel availability check (CAC) and minimize channel switches caused by false radar detection events. Table 7.2 shows the number of nonoverlapping channels in Federal Communications Commission

(FCC) and European Telecommunications Standards Institute (ETSI) regulatory domains for different channel bandwidth. A practical problem worth mentioning is that older client devices do not support channel 144, so the majority of AP deployments disable this channel, which results in removal of one 20-MHz, one 40-MHz channel and one 80-MHz channel.

**Table 7.2**

Number of Nonoverlapping 5-GHz channels in FCC and ETSI Domain

| *Bandwidth (MHz)* | *DFS Support* | *Number of Channels* | |
|---|---|---|---|
| | | *FCC Domain* | *ETSI Domain* |
| 160 | Enabled | 2 | 2 |
| 160 | Disabled | 0 | 0 |
| 80 | Enabled | 6 | 5 |
| 80 | Disabled | 2 | 1 |
| 40 | Enabled | 12 | 10 |
| 40 | Disabled | 4 | 2 |
| 20 | Enabled | 25 | 20 |
| 20 | Disabled | 9 | 4 |

If the number of APs in a network exceeds the number of available nonoverlapping channels, then some AP has to reuse a channel at a different location. This is called channel reuse or frequency reuse and this method is used to scale the network capacity with increasing number of APs. The least distance between any two APs or cells in a network that operate on the same channel is known as frequency reuse distance. To avoid cochannel interference or self-interference in a network, the frequency reuse distance has to be maximized. However, the number of available nonoverlapping channels imposes upper bounds on the achievable frequency reuse distance. One of the methods to arrive at a channel plan that maximizes frequency reuse distance is to first construct a cluster of cells with nonoverlapping channels assigned to each cell and then simply repeating this cell cluster pattern to cover the entire venue. Such repeatable cell cluster patterns that can cover any given area without gaps can be constructed [1] for any cluster size $N = i^2 + i \cdot j + j^2$ where $i, j$ are any integers. It can also be shown that the frequency reuse distance $D$ for such a channel plan is given by $D = r\sqrt{3 \cdot N}$ where, $r$ is the cell radius. As an example, for $i = 1, j = 1$, the cluster size $N = 3$ and this three cell cluster can be used in 2.4 GHz, which has three nonoverlapping channels (1, 6 and 11). Figure 7.3 depicts the cell cluster patterns for common cluster sizes of 3, 4, and 7.

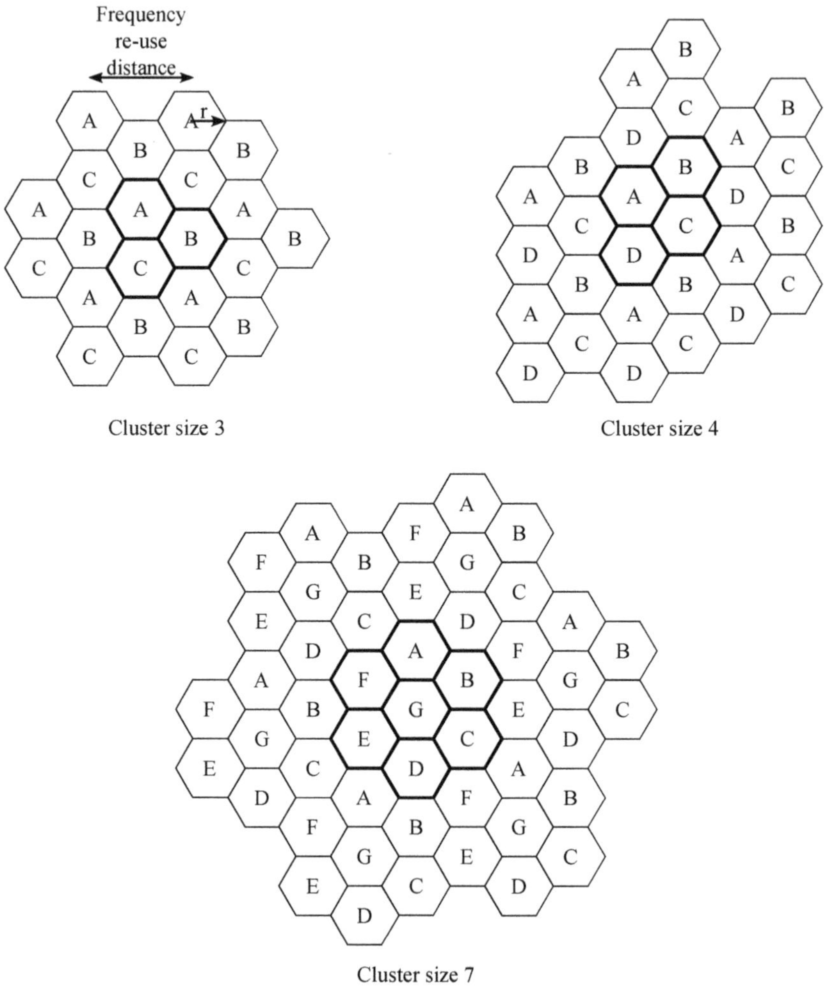


**Figure 7.3** Three, four, and seven cell clusters.

In summary, a smaller channel bandwidth configuration opens up more available channels, which permits a larger cluster size minimizing the number of cochannel neighbor APs. However, this comes at the expense of downgrading the peak throughput of an individual AP. The optimal channel bandwidth setting depends on the density of AP deployment and the acceptable number of cochannel neighbor APs.

### 7.2.4 Estimating the Number of APs

One of the key challenges in deployment planning is determining the density of AP deployment. There are two important considerations to determine the number of APs required for an enterprise deployment venue. One is coverage and the other is capacity. Coverage concerns the ability to provide a minimum level of receive signal strength indicator (RSSI) to a client located anywhere across the venue. This determines the minimum downlink modulation and coding scheme (MCS) or physical (PHY) rate a client would experience in the venue. Capacity, on the other hand, concerns the ability of the network to meet the per-client throughput demands when multiple clients simultaneously have active traffic. Underdeployment of APs results in weak RSSI spots or dead spots with no connectivity and failure to meet the throughput demands as the client count increases in the network. An overdeployment of APs results in increased cost, higher airtime wasted on management traffic and can cause higher cochannel interference between cells. Good planning for both coverage and capacity can help avoid issues of underdeployment and overdeployment.

#### 7.2.4.1 Planning for Coverage

The minimum RSSI required in a venue does depend on the application needs. As an example, a warehouse using Wi-Fi for asset tracking requires only basic connectivity, so a minimum RSSI of -82 dBm providing MCS 0 PHY rate is good enough, whereas, applications demanding quality of service (QoS) such as video surveillance, video conferencing, video streaming, and Voice over IP (VoIP) require higher data rates and may need a minimum RSSI of -60 dBm. Once the minimum RSSI requirement is finalized, the physical cell size of one AP can be determined by calculating the distance from AP that will result in minimum RSSI. As illustrated in Figure 7.4, if the RSSI at the point of intersection of two adjacent cells equals the minimum RSSI, then the RSSI at any location will exceed the minimum RSSI.

To calculate the distance from AP that will result in minimum RSSI, it is required to have knowledge of the AP Tx power and the average path loss exponent for the venue. The path loss exponent depends on the type of building material used, the number of walls, pillars, and other internal structures that have an impact on RF propagation. The path loss exponent can be accurately obtained

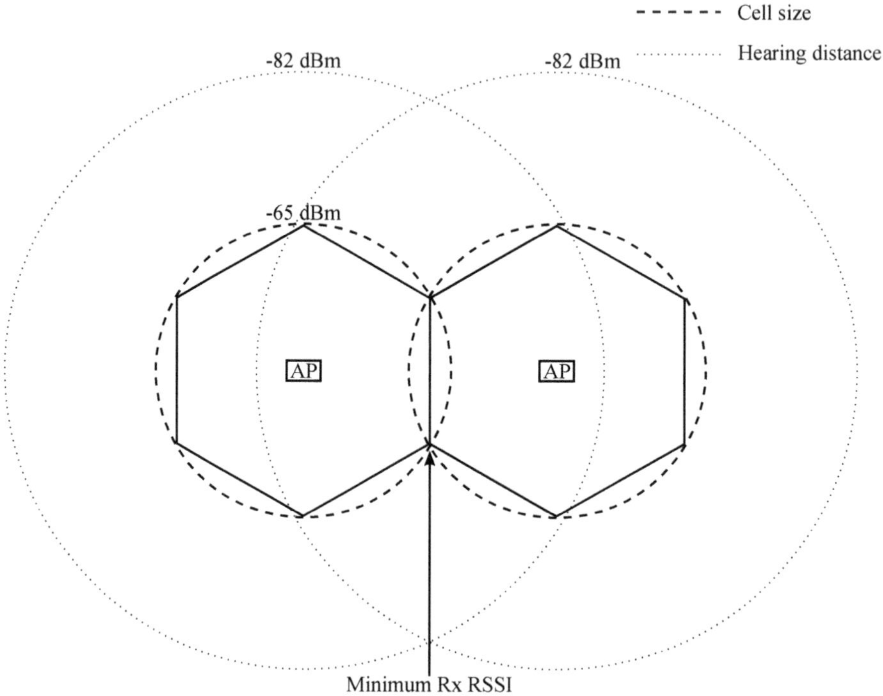


**Figure 7.4** Minimum RSSI at intersection of two cells.

from a site survey measurement using a single AP and client or it can be assumed based on knowledge of the venue environment. Since the path loss increases with channel frequency as $20 \cdot \log_{10}$(channel frequency) and the Tx power of AP might vary with channel due to regulatory limits, the channel with the lowest AP Tx power $- 20 \cdot \log_{10}$(channel frequency) should be considered for cell size determination. Considering uplink traffic though, the PHY rate depends on the client's Tx power, which is not in the control of deployment. So in practice, the channel with the highest frequency is typically used to determine the cell size. For example, if it is a triband Wi-Fi 6E AP deployment in the United States, then channel 233 (7115 MHz) should be used for coverage planning purpose. To give a sense of path loss disparity, the path loss in the 5-GHz band is approximately 7 dB higher than the 2.4-GHz band, and the path loss in the 6-GHz band is approximately 2 dB higher than the 5-GHz band. In addition, the material absorption loss of walls and other internal structures is higher in the 5-GHz band as compared to the 2.4-GHz band. Due to this path loss and material absorption loss disparity, the Tx power of multiband capable APs has to be reduced in the lower frequency bands in order to have the same cell overlap area across frequency bands. Otherwise, the number of AP neighbors in the 2.4-GHz band will be higher relative to the 5-GHz band leading to more inter-AP interference in the 2.4-GHz band. Moreover, the clients will always see a higher RSSI in the 2.4-GHz band and this could influence clients to prefer the 2.4-GHz band over the relatively cleaner 5-GHz band, which has more available channels. This is one of the common Wi-Fi deployment issues that leads to worse network performance.

Another consideration for coverage planning in an enterprise network is to ensure a client can hear two APs at all intended coverage areas. This will guarantee a client has a backup AP to roam in case one AP either changes channel to a DFS channel and performs a CAC or fails for any reason such as out of memory, software crash or reboot. Finally, the number of APs required to meet the coverage requirement can be calculated by dividing the area of the venue by the cell area. Good coverage planning ensures a minimum downlink and uplink PHY rate for a client anywhere across the venue, but it ignores the performance scaling aspect with multiple clients, which is addressed by capacity planning.

#### 7.2.4.2 Planning for Capacity

The objective of capacity planning in a deployment is to estimate the number of APs required to ensure the network can handle the peak client load and traffic load that is expected during network operation. The various factors involved in capacity planning are listed below:

1. Capabilities of AP such as number of frequency bands supported, Wi-Fi generation of AP, and number of Tx and Rx antennas.
2. Peak number of clients with concurrently active traffic expected in the network.
3. Different types of client capabilities and the expected number of clients belonging to each type. For example, estimated number of Wi-Fi 6E clients, 2.4-GHz only clients and Wi-Fi 4 clients.
4. Amount of application traffic running on different client types.

As it is difficult to do a precise estimation considering all the aspects, typically several approximations are made to simplify the capacity planning. Let us see a detailed worked-out example to understand the various approximations and steps involved in capacity planning. For this example, let us assume the APs are 2 × 2 tri-band Wi-Fi 6E capable and configured to a channel bandwidth of 20 MHz in the 2.4-GHz band, 80 MHz in the 5-GHz band, and 160 MHz in the 6-GHz band. Also, coverage planning is assumed to be completed with a minimum RSSI requirement of -65 dBm. Since different client types with different capabilities connect to a network, the first step in capacity planning is to estimate the application layer TCP throughput achievable at minimum RSSI for different client types assuming 100% airtime is available. From the minimum RSSI, the minimum PHY rate can be estimated using the IEEE Rx sensitivity requirement (see Table 5.3). Then, the achievable transmission control protocol (TCP) throughput is approximated as 65% of PHY rate to factor in retransmission overhead due to packet error rate (PER) and higher-layer protocol overheads such as TCP header, IP header and TCP ACK. Since the AP channel bandwidth varies with the frequency band, a client will experience different throughput in different frequency bands, as detailed in Table 7.3.

**Table 7.3**

TCP Throughput Achievable at -65 dBm RSSI

| *Client* | *Band* | *BW (MHz)* | *MCS* | *PHY Rate (Mbps)* | *TCP Throughput (Mbps)* |
|---|---|---|---|---|---|
| Wi-Fi 1 | 2.4 GHz | 20 | | 11 | 7.1 |
| Wi-Fi 2, 3 | 2.4 GHz | 20 | | 54 | 35.1 |
| | 5 GHz | 20 | | 54 | 35.1 |
| 1 × 1 Wi-Fi 4 | 2.4 GHz | 20 | 6 | 65 | 42.2 |
| | 5 GHz | 40 | 5 | 120 | 78 |
| 2 × 2 Wi-Fi 5 | 2.4 GHz | 20 | 6 | 130 | 84.5 |
| | 5 GHz | 80 | 4 | 390 | 253.5 |
| 3 × 3 Wi-Fi 5 | 2.4 GHz | 20 | 6 | 195 | 126.7 |
| | 5 GHz | 80 | 4 | 585 | 380.2 |
| 2 × 2 Wi-Fi 6 | 2.4 GHz | 20 | 6 | 154.8 | 100.6 |
| | 5 GHz | 80 | 4 | 432.4 | 281 |
| 2 × 2 Wi-Fi 6E | 2.4 GHz | 20 | 6 | 154.8 | 100.6 |
| | 5 GHz | 80 | 4 | 432.4 | 281 |
| | 6 GHz | 160 | 3 | 576.4 | 374.6 |

The next step in capacity planning is to estimate the peak number of client devices that have concurrently active traffic and also the amount of application traffic running on these devices concurrently. For illustration purpose, let us assume the following list of concurrently active traffic on different client types:

1. Twenty-five Wi-Fi 1 capable printers with a per-client throughput of 1 Mbps.
2. Ten Wi-Fi 2,3 capable Voice over IP (VoIP) phones with a per-client throughput of 300 kbps.
3. One hundred $1 \times 1$ Wi-Fi 4 smartphones with a per-client throughput of 5 Mbps.
4. Two hundred $2 \times 2$ Wi-Fi 5 laptops with a per-client throughput of 5 Mbps.
5. Fifty $3 \times 3$ Wi-Fi 5 laptops and desktops with a per-client throughput of 10 Mbps.
6. Fifty $2 \times 2$ Wi-Fi 6 smartphones with a per-client throughput of 10 Mbps.
7. Five $2 \times 2$ Wi-Fi 6E smart televisions with a per-client throughput of 50 Mbps.

To simplify the capacity analysis, let us assume the clients are uniformly distributed across the venue area and that each of them experience a RSSI of -65 dBm. Owing to the coverage planning, most clients are expected to observe a RSSI greater than -65 dBm, therefore this simplifying assumption on RSSI provides a worst-case analysis. Now, the airtime required to meet the above per-client throughput can be roughly estimated using Table 7.3. For example, Table 7.3 shows a peak throughput of 7.1 Mbps for a Wi-Fi 1 client device at -65 dBm RSSI. Hence, twenty-five Wi-Fi 1 devices with a per-client throughput requirement of 1 Mbps would require an airtime of $\frac{25 \cdot 1}{7.1} = 352\%$ relative to a single AP's theoretical airtime capacity. This implies at least four APs are required to meet this airtime requirement in the 2.4-GHz band. In practice, $100\%$ of the airtime is not entirely available for data transfer as some fraction of airtime is wasted due to management frame overhead, collisions, random backoff (RBO), retransmissions, and non-Wi-Fi interference. Moreover, the medium has to be shared with other APs and clients within hearing range on the same channel. For the purpose of this example, let us assume the cochannel interference between APs is mitigated with proper channel planning, although it is understood that there are practical limitations especially in 2.4-GHz. Therefore, the available airtime on a AP must be assumed to be much lower than $100\%$ to accommodate these overheads and also to reserve some airtime for future capacity expansion. For the purpose of this example, let us assume the available airtime per AP is 50%. Hence, to serve the throughput needs of the twenty-five Wi-Fi 1 devices alone, $\frac{352}{50} \approx 7$ APs are required. Similarly the percentage of airtime required for applications running on each client type can be estimated as shown in Table 7.4. For dual-band capable client types, let us assume the ratio of the number of clients connected in 2.4-GHz band to that in the 5-GHz band is 1:4, which aligns with the channel bandwidth ratio of $20 : 80 = 1 : 4$ between the 2.4-GHz and 5-GHz bands. For a triband capable client type, let us assume 100% of such clients are connected in the 6-GHz band. Summing up the airtime requirement in each frequency band, the number of APs required to meet the capacity turns out to be 22 for this specific example, as detailed in Table 7.4.

**Table 7.4**

Capacity Planning Example.

| *Client Type* | *Per-client Throughput (Mbps)* | *Number of Active Devices* | | | | *Airtime Required* | | |
|---|---|---|---|---|---|---|---|---|
| | | *Total* | *2.4 GHz* | *5 GHz* | *6 GHz* | *2.4 GHz* | *5 GHz* | *6 GHz* |
| Wi-Fi 1 | 1 | 25 | 25 | 0 | 0 | 352% | 0% | 0% |
| Wi-Fi 2,3 | 0.3 | 10 | 2 | 8 | 0 | $\frac{2 \cdot 0.3}{35.1} \approx 1.7\%$ | $\frac{8 \cdot 0.3}{35.1} \approx 6.8\%$ | 0% |
| $1 \times 1$ Wi-Fi 4 | 5 | 100 | 20 | 80 | 0 | 237% | 513% | 0% |
| $2 \times 2$ Wi-Fi 5 | 5 | 200 | 40 | 160 | 0 | 236% | 316% | 0% |
| $3 \times 3$ Wi-Fi 5 | 10 | 50 | 10 | 40 | 0 | 79% | 105% | 0% |
| $2 \times 2$ Wi-Fi 6 | 10 | 50 | 10 | 40 | 0 | 99.4% | 143% | 0% |
| $2 \times 2$ Wi-Fi 6E | 50 | 5 | 0 | 0 | 5 | 0% | 0% | 66.7% |
| Total airtime requirement | | | | | | 1005% | 1084% | 66.7% |

Total number of APs needed in 2.4-GHz $= \frac{1005}{50} \approx 21$

Total number of APs needed in 5 GHz $= \frac{1084}{50} \approx 22$

Total number of APs needed in 6 GHz $= \frac{66.7}{50} \approx 2$

Finally, the total number of APs required is chosen to be the maximum of both coverage planning and capacity planning exercise. While for simplicity reasons, it was assumed in coverage planning that the clients are uniformly distributed across the venue, this is often not true in practice as the client density varies across different parts of the venue over time. For example, in an office or university, people gather at the cafeteria or break room during lunch-time. As another example, during an office event, people may gather in a large conference hall to hear an important announcement. So, in certain areas of the venue where client density can occasionally peak, more APs must be deployed considering the peak load in that specific area.

### 7.2.5 Mitigating Inter-AP Interference

In some deployment scenarios, the number of APs required for meeting capacity demands exceeds the coverage demands. This is common in high client density public venues having high throughput requirements, such as stadiums for example. In such cases, depending on the density of APs deployed, multiple neighbor APs may be within the hearing distance of an AP, as depicted in Figure 7.4. Hearing distance is the maximum distance from the AP up to which an AP can receive Wi-Fi frames from a neighbor AP. Note that the hearing distance of an AP is much larger than the cell size. As long as the neighbor APs are on nonoverlapping channels, inter-AP interference or self-interference is not an issue. In other words, the frequency reuse distance has to be maintained higher than the hearing distance to mitigate inter-AP interference. In most scenarios, the number of cochannel neighbors can be kept under control with good channel planning. However, in some high-density AP deployments, the number of cochannel neighbors may be large enough to cause significant inter-AP interference and this can severely limit the capacity of the network. Some common methods to mitigate inter-AP cochannel inference are listed below:

1. Increase the number of available channels by reducing channel bandwidth. This is one of the simplest methods to reduce inter-AP interference but it has the downside of bringing down the peak capacity of every AP. Some network administrators don't like this solution as they feel they are underutilizing an AP and not actually getting what they paid for. While this perspective is understandable, from a network capacity point of view this may well be a good solution.

2. Reducing the hearing distance by reducing the maximum Tx power or increasing the minimum sensitivity of AP. This is an elegant solution to the problem but the amount of Tx power reduction or minimum sensitivity increase has to be carefully decided. If it is either overdone or unnecessarily done, it can lead to weak spots in coverage.

3. Increase the management frame data rate of AP and minimum data rate of BSS advertised to clients. Since a higher data rate requires higher Rx sensitivity, this effectively reduces the hearing distance.

4. Use the spatial reuse feature in Wi-Fi 6. However, this feature is useful only with Wi-Fi 6 clients and therefore relevant only in deployments with a high percentage of Wi-Fi 6 clients. For example, this can be useful in the 6-GHz band, but since there are more channels in 6 GHz, it may not be a problem today.

### 7.2.6 Upgrading Existing Deployment

When upgrading an existing deployment with the latest AP model, it is recommended to repeat the site survey and coverage planning exercise. Some aspects, if overlooked, can result in poor network performance after the network upgrade contrary to expectations. Below are some key points to take note of during AP upgrade in an existing deployment:

1. Compensate for any differences in equivalent isotropically radiated power (EIRP) and Rx sensitivity between the old and new AP models in all applicable frequency bands. If the newer AP model has a higher (EIRP - Rx sensitivity) metric, the hearing distance increases leading to potentially more cochannel neighbors. On the other hand, if the newer AP model has lower (EIRP - Rx sensitivity) metric, it leads to reduction of minimum RSSI. This issue is especially relevant when the number of Tx, Rx antennas of the newer AP is different. When performing a one-to-one swap of APs, the Tx power or Rx sensitivity of the newer AP model has to be adjusted to compensate for any difference in (EIRP - Rx sensitivity) metric.

2. If support for a new frequency band is added, then the coverage planning exercise should be repeated to ensure good coverage in the new frequency band. For example, when upgrading from 2.4-GHz only Wi-Fi 3 to triband Wi-Fi 6E, some APs might have to be added in the 6-GHz band to provide the same coverage as in the 2.4-GHz band.

3. Compare the power consumption requirement of the old and new AP models and upgrade the PoE network switch and Ethernet cabling as applicable. For example, triband Wi-Fi 6E APs typically draw more power owing to concurrent operation of three radios and therefore require higher PoE power like 802.3bt to provide optimal performance. If 802.3at power is provided to an AP requiring 802.3bt power, the AP may still work by reducing its Tx power, downgrading the number of Tx antennas, or turning off some interfaces to keep its power consumption below 25 W, but the AP cannot deliver its best performance.

4. Upgrade the network switch and Ethernet cabling to match or exceed the Ethernet port speed capability of the newer AP model.

### 7.2.7 Reducing Overhead in High-Density Deployments

As the density of deployment increases, the management frame overhead also increases, which eats into the data airtime. Below are some common techniques to reduce overhead in high-density deployments.

1. Keep the number of virtual APs (VAPs) or basic service set (BSS) to a minimum.
2. Enable enhanced multi-BSSID advertisement (EMA) mode of operation with multiple BSS when all clients support EMA such as in the 6-GHz band.
3. Increase the management frame data rate of AP and minimum data rate of BSS advertised to clients. This reduces the airtime occupied by management frames and also prevents association of clients to far away APs.
4. If there are no Wi-Fi 1 clients expected in the network, disable support for Wi-Fi 1. This can help reduce overheads in the 2.4-GHz band.
5. Use proxy address resolution protocol (ARP) service [2] on the AP to reduce broadcast ARP frames on the wireless medium.

## 7.3 OUTDOOR WI-FI DEPLOYMENT

Outdoor Wi-Fi deployment planning is similar to indoor deployment planning but with several additional constraints. A major difference is the lower path loss exponent in an outdoor environment, which is close to the free space path loss model. Hence, the cell size and hearing distance are relatively much larger in an outdoor environment. There are several physical and mechanical constraints that limit the flexibility in placement of APs. Installing a wired backhaul link may not be possible in certain locations, so wireless mesh has to be used in such locations. As well, weather-proof enclosures are necessary to protect the AP from nature's harsh elements. Outdoor APs typically come with different options for the antenna such as omnidirectional, 120-degree sectored, and narrowbeam, as shown in Figure 7.5.

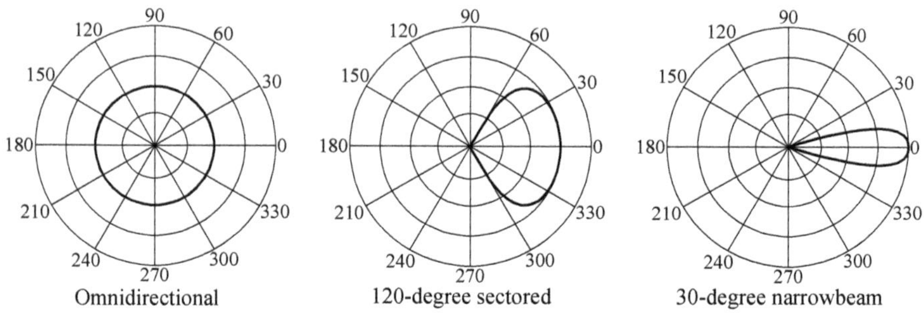


**Figure 7.5** Horizontal plane pattern of omnidirectional, sectored, and narrowbeam antennas.

While sectored and narrowbeam antennas do not provide additional coverage compared to omnidirectional antennas due to regulatory EIRP limitation, they are useful for limiting self-interference in a network. For example, in an outdoor stadium with hundreds of deployed APs, sectored or narrowbeam antennas can be used to limit the coverage of each AP to a small section of the stadium, thereby limiting interference to other APs in the stadium. Sectored antennas can also enable capacity increase by limiting the cell area to a certain sector and employing multiple APs with frequency reuse to cover all sectors. Another category of outdoor APs is point-to-point AP products for fixed wireless applications. In such point-to-point AP products, highly directional antennas are mounted on top of buildings to provide several kilometers of range between two fixed-point locations using direct line of sight. There are some channels allotted in FCC and ETSI domain that permit significantly higher EIRP limit for such point-to-point applications. However, due to the absence of rich multipath fading in direct line-of-sight transmission, spatial multiplexing is hard to achieve. Nevertheless, up to two spatial stream multiple input multiple output (MIMO) has been achieved by several point-to-point AP products using a combination of vertical and horizontal polarization antennas for uncorrelated signals.

## 7.4 POST DEPLOYMENT SITE SURVEY

After the deployment is completed, it is recommended to perform a site survey taking measurements at multiple locations with a client in sniffer mode. In sniffer mode, all frames received by a client are captured and stored. At each location, the beacons received from different BSS in each channel are recorded along with the beacon RSSI and BSS identifier (BSSID) information. By sampling enough locations, the minimum RSSI in the network, the actual cell size, and hearing distance of each AP, the overlap area between cells and the cochannel neighbor APs can be determined. This post-deployment exercise helps assess and confirm whether the coverage planning exercise really achieved its objective. In addition, a throughput test can be run at different locations to spot check the performance health of APs.

## 7.5 REFERENCES


[1] Bruce Alexander, *802.11 Wireless Network Site Surveying and Installation*, Cisco Press, 2004.

[2] "IEEE Standard for Information Technology–Telecommunications and Information Exchange between Systems - Local and Metropolitan Area Networks–Specific Requirements - Part 11: Wireless LAN Medium Access Control (MAC) and Physical Layer (PHY) Specifications," *IEEE Std 802.11-2020 (Revision of IEEE Std 802.11-2016)* (2021), pp. 1–4379, DOI: 10.1109/IEEESTD.2021.9363693.

# Chapter 8

## 5G Overview and Wi-Fi Comparison

Many different wireless technologies have been developed over the last four decades, some in direct competition with one another and several others to serve a specific use case. These technologies are commonly divided into four broad categories based on data rate and coverage range:

- Wireless personal area network (WPAN) technologies are intended for low-power, short-range communication between devices, such as smartphones and wireless headsets or laptops and wireless keyboards, controlled by a single person. The reach of WPAN varies from a few centimeters to a few meters. IrDA and Bluetooth are two common WPAN examples.
- Wireless sensor network (WSN) technologies are designed to interconnect low-power, low-cost sensors that monitor physical and environmental conditions such as temperature, pressure, or pollutants, and cooperatively pass the data to a central location, where it can be observed and analyzed. Zigbee, Bluetooth Low Energy (BLE), and WirelessHART are some of the widely used WSN technologies.
- Wireless local area network (WLAN) technologies provide high data rate untethered final link over several tens of meters between the existing wired network and client devices enabling wireless access to other systems on the local network or the Internet. Wi-Fi and HIPERLAN are examples of WLAN technologies although HIPERLAN is no longer in widespread use.
- Wireless wide area network (WWAN) technologies enable communication across hundreds of meters. They rely on a network of base stations deployed across a vast coverage area. Each base station serves users in a part of the

coverage area referred to as a cell, hence WWAN technologies are more commonly referred to as cellular communication technologies. Over the last 40 years, the world has witnessed a new generation of cellular communication technologies every decade, ranging from Advanced Mobile Phone System (AMPS) to Long Term Evolution (LTE). While AMPS is considered a first-generation technology, LTE belongs to the fourth generation. 5G is the fifth and the latest generation cellular technology.

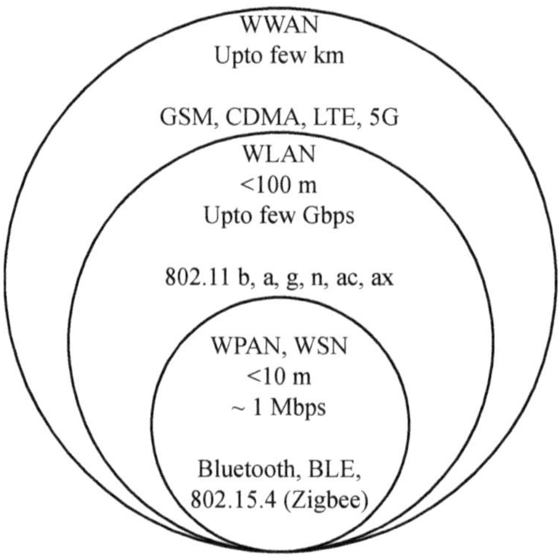


**Figure 8.1** Wireless network technologies. (*Source:* Nanocell Networks Pvt. Ltd.)

Figure 8.1 summarizes the salient aspects of the four categories of wireless network technologies. Although each of these technologies are designed with certain unique attributes to address specific needs, sometimes there are overlapping use cases that often lead to a debate on whether one technology can totally replace the other. The launch of 5G has further escalated this debate especially given its aspiration to address local area use cases in addition to the traditional wide area use cases. A book on Wi-Fi would be remiss not to include a discussion on 5G and on how it compares to Wi-Fi 6 so that the reader is equipped with the information required to determine which technology best suits their needs.

## 8.1 EVOLUTION OF CELLULAR TECHNOLOGIES (1G-4G)

Just like Wi-Fi, cellular technologies have evolved through multiple generations [1]. In fact, the idea of classifying technologies by generation originated in the cellular world and was later adopted by Wi-Fi in 2018 around the time products based on the sixth-generation technology were getting launched. While each Wi-Fi generation maps to only one technology, each cellular generation consists of multiple technologies that are incompatible.

### 8.1.1 First Generation Cellular Technologies

The first generation technologies (1G), which emerged around 1980, were based on analog transmission that used frequency modulation (FM), frequency division duplexing (FDD), and frequency division multiple access (FDMA). Being analog, 1G was inherently insecure and was limited to voice services. The more successful 1G technologies were AMPS in the United States and its variant, total access communication systems (ETACS and NTACS) in Europe and Japan. These were almost identical from a radio standpoint, with the major difference being the channel bandwidth. The AMPS system used 30-kHz channel bandwidth, whereas ETACS and NTACS used 25 kHz and 12.5 kHz, respectively. Another difference between AMPS and ETAC is how the telephone number of each subscriber is formatted, due to the need to accommodate different country codes throughout Europe as opposed to area codes in United States. Due to these differences, AMPS, ETACS, and NTACS are incompatible with each other although all three are considered first-generation cellular technologies. This trend of multiple incompatible technologies within each generation continued all the way to the fourth generation. Furthermore, there is no cross-generational compatibility either, which is very different than forward and backward compatibilities guaranteed in the Wi-Fi world.

Most of the early systems were designed for a carrier-to-interference ratio (CIR) of 18 dB for satisfactory voice quality and were deployed in a 7-cell frequency reuse pattern with 3 sectors per cell.

### 8.1.2 Second Generation Cellular Technologies

The first major upgrade to cellular communication arrived in the early 1990s with the introduction of second generation (2G) technologies based on digital transmission. The target service was still voice although the use of digital transmission allowed 2G to support limited data services. Adoption of digital modulation techniques, along with multiple access techniques such as time division multiple access (TDMA) and code division multiple access (CDMA), enabled 2G to achieve three times greater system capacity compared to 1G. Furthermore, the required CIR to meet satisfactory voice quality dropped from 18 dB to just a few dB due in large part to error correction coding and digital signal processing techniques used in 2G systems. Simple encryption was also introduced in 2G as a measure of security against eavesdropping and fraud.

Examples of 2G cellular technologies include the global system for mobile communications (GSM), IS-95 CDMA, and IS-136 TDMA systems. GSM is by far the most widely deployed of these. The personal handy-phone system (PHS) deployed in Japan, China, and some other Asian countries is also considered a 2G system.

In addition to providing improved capacity and security, 2G technologies enabled new data applications, short messaging service (SMS) being the prime one among them. Besides SMS, 2G systems also supported low data rate wireless data services such as the delivery of news, stock quotes, and weather. Original 2G systems relied on circuit switching for data services, but later evolved to support packet switching as well. Limitations in data rate and available space for display in handheld devices led to the development of specialized protocols, such as WAP, to tailor and deliver internet content to handheld devices.

### 8.1.3 Third Generation Cellular Technologies

The third generation (3G) was introduced in 2001 to facilitate greater voice and data capacity, thereby laying the foundations for mobile broadband. 3G technologies were a significant leap over 2G providing much higher data rates. Besides higher data rate, 3G systems also targeted better quality of service (QoS) control tailored for a variety of applications ranging from voice telephony and interactive games to web browsing, e-mail, and streaming multimedia applications.

A collaborative project, called the 3rd generation partnership project (3GPP), was established in the late 1990s by bringing together several telecommunications standards bodies with the initial goal of developing a new technology for the third-generation cellular networks as an evolution of the TDMA-based 2G GSM technology. Around the same time, a parallel group called 3rd generation partnership project 2 (3GPP2) was formed to develop an alternative 3G technology as an evolution of the CDMA-based 2G IS-95. Ultimately, both 3GPP and 3GPP2 converged towards using CDMA as the underlying baseline technology for 3G standards although some differences remained. While the 3G technologies developed by 3GPP were called wideband CDMA (W-CDMA) or universal mobile telecommunication system (UMTS) and utilized a 5-MHz bandwidth signaling, 3GPP2 technologies were called cdma2000 and utilized a 1.5-MHz bandwidth signaling. Both W-CDMA and cdma2000 were eventually endorsed as 3G technologies.

The scope of 3GPP was later extended to encompass the development and maintenance of technologies for the 4th generation and 5th generation cellular networks. 3GPP2, on the other hand, has mostly gone dormant and exists just to maintain cdma2000.

### 8.1.4 Fourth Generation Cellular Technologies

Motivated by the ever-increasing demand for mobile broadband services with higher data rates and better QoS, work began on the fourth generation (4G) technologies. The requirements for 4G included not only peak data rates exceeding 100 Mbps, but also lower network latency and an all-IP packet switched network that uses IP even for voice. Three competing technologies emerged as potential solutions for 4G: long term evolution (LTE) from 3GPP, ultra mobile broadband (UMB) from 3GPP2, and WiMAX from IEEE. Ultimately, LTE became the most predominantly adopted and deployed 4G technology, and has since become the global standard for 4G.

LTE introduced several new technologies—orthogonal frequency division multiplexing (OFDM), multiple input multiple output (MIMO), and system architecture evolution (SAE)—that were not present in the earlier generation cellular systems. OFDM and MIMO enable LTE to utilize the spectrum more efficiently and achieve much higher data rates. SAE is the network evolution brought in to meet the very high data rate and low latency requirements. One key change is the shifting of several functions previously handled by the core network to the edge. Essentially this provides a flatter network architecture resulting in reduced latency and enabling more direct routing of data to its destination.

Besides very high-speed mobile internet access, 4G enabled applications such as video conferencing, gaming services, IP telephony, and high-definition (HD) mobile TV.

## 8.2 FIFTH GENERATION CELLULAR TECHNOLOGY

5G is the fifth and the latest generation of cellular technology that brings in three key capabilities [2]:

- *Wider channels* to achieve faster speeds and high capacity for enhanced mobile broadband (eMBB) applications,
- *Lower latency* to be more responsive for ultra reliable low latency communications (URLLC) and
- *Ability to concurrently connect a lot more devices* for massive machine type communications (mMTC) used by sensors and smart devices.

3GPP has developed a new radio-access technology called NR (new radio) for 5G cellular networks. As is typical with cellular technologies, NR is not backward-compatible with any of the earlier generation technologies including 4G LTE, although it is expected to be forward-compatible with future generation cellular technologies. To meet the increased throughput demands of 5G, in parallel to NR, 3GPP has developed a new 5G core network referred to as 5GCN. The core network is a central part of the cellular network. It makes it possible for subscribers to get access to the services they are entitled to by providing functions including authentication, security, session management, and aggregation of traffic from end devices. The new 5G radio-access technology will connect to the 5GCN.

Besides 5GCN, NR may also connect to evolved packet core (EPC), which is the legacy core network used by 4G LTE. This makes it easier for network operators to deploy 5G networks leveraging the existing 4G infrastructure. This mode of operation is called nonstandalone access (NSA) since NR devices rely on LTE for initial access and mobility as illustrated in Figure 8.2. However, standalone access (SA), where NR connects to 5GCN and uses 5G for both signaling and data transfer without the need for an existing LTE network, is the true 5G network capable of achieving high data rates and minimal network latency. The differences between SA and NSA affects higher layers and the interface to the core network as the underlying radio technology remains the same in both cases.

The initial 5G specification (also known as Release 15) published in December 2017 is limited to NSA mode, while the latest specification (Release 16) finalized in July 2020 supports SA mode as well. Devices that support both NSA and SA modes are referred to as dual-mode devices. Since there is still no standard for voice calls over 5G, phones must fall back to 4G whenever a phone call is made. The 5G specification also allows phones to aggregate 5G and 4G channels seamlessly and invisibly to the user.

In order to accelerate installation times and to eliminate the need to invest in expensive hardware, 5G employs network function virtualization (NFV) to decouple software from hardware by replacing various network functions such as firewalls, load balancers, and routers with virtualized instances running as software.

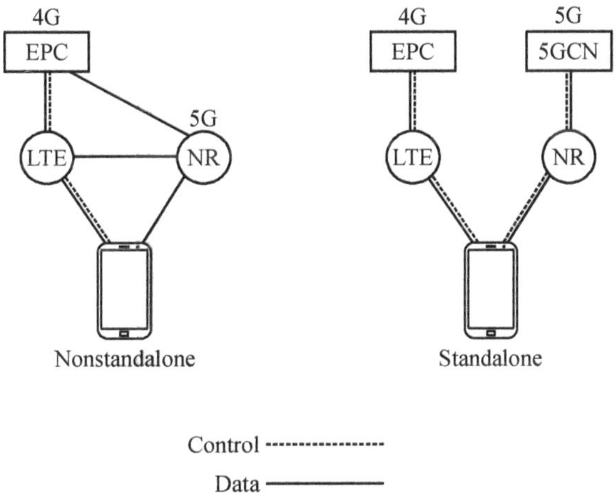


**Figure 8.2** 5G nonstandalone and standalone access.

Unlike Wi-Fi 6 that is restricted to operate in the unlicensed spectrum in 2.4-GHz, 5-GHz, and the recently allocated 6-GHz bands, NR can be deployed in licensed spectrum from below 1 GHz up to 52.6 GHz as well as in the new unlicensed 6-GHz band. The NR specification refers to the frequency range 0.45 – 6 GHz as frequency range 1 (FR1), while the range 24.25 – 52.6 GHz is referred to as FR2. FR1 is commonly known as sub-6 GHz, while FR2 is commonly known as millimeter wave (mmWave) although the millimeter wavelength starts at 30 GHz.

While operation at mmWave frequencies enables high data rates and large capacity due to the availability of large amounts of contiguous spectrum, its coverage range is constrained by the higher propagation loss at higher frequencies. The array gain achievable by leveraging massive array of mmWave antennas, which can be fitted into small form factors due to their much smaller size compared to sub-6-GHz antennas, helps compensate for the higher propagation loss. However, this gain is insufficient to meet the link budget requirements in scenarios that entail non-line-of-sight or outdoor-to-indoor propagation. This is further complicated by the more stringent regulatory limits on the maximum permissible exposure at mmWave frequencies especially for handheld devices. For these reasons, sub-6-GHz operation will continue to prevail even in the 5G era.

Similar to Wi-Fi 6, NR uses OFDM signaling owing to its ability to cope with severe channel conditions without complex equalization filters. However, unlike

Wi-Fi 6, NR supports conventional non-DFT-precoded OFDM as well as DFT-precoded OFDM in the uplink. The latter, not supported in Wi-Fi 6, helps reduce the peak-to-average power ratio (PAPR) problem of OFDM, thereby improving the power amplifier efficiency on the device side at the cost of increased uplink receiver complexity.

Wi-Fi 6 supports 312.5-kHz and 78.125-kHz subcarrier spacings that are optimized for indoor and outdoor deployments, respectively. On the other hand, to support a wide range of deployment scenarios, from macro cells with limited bandwidth in sub-6-GHz spectrum to small cells with very wide bandwidth in the mmWave spectrum, NR supports subcarrier spacings ranging from 15 kHz to 120 kHz. Smaller subcarrier spacing has the benefit of providing relatively longer multipath tolerance at a reasonable cyclic prefix overhead, while larger subcarrier spacings are needed to handle the increased phase noise at higher carrier frequencies. While Wi-Fi 6 supports up to 2048 OFDM tones and signaling bandwidths of 20/40/80/160 MHz, NR supports up to 3300 tones with 50/100/200/400 MHz signaling bandwidths. Furthermore, NR allows use of carrier aggregation to support bandwidths larger than 400 MHz.

Cellular spectrum allocation often tends to be paired (two distinct bands, one assigned to downlink and other to uplink) in lower frequency bands and unpaired (single band used for both downlink and uplink) in higher frequency bands. In order to support paired and unpaired spectrum allocations, NR supports both time division duplex (TDD) and frequency division duplex (FDD), while Wi-Fi 6 limits duplex to TDD as the unlicensed Wi-Fi spectrum is unpaired.

Both NR and Wi-Fi 6 use low density parity check (LDPC) codes for error correction coding. In addition to LDPC, Wi-Fi 6 supports the option of using binary convolutional codes (BCC) as well to maintain backward compatibility. While 1024-QAM is the maximum constellation supported in Wi-Fi 6, it is 256-QAM in the case of 5G NR.

NR supports both analog and digital beamforming. As the name suggests, digital beamforming is done in the digital domain whereas analog beamforming is done in the analog domain. Analog beamforming is better suited for mmWave operation that uses a massive antenna array, while digital beamforming works better for sub-6-GHz operation with relatively fewer antennas. Since Wi-Fi 6 operation is restricted to low frequency bands, it supports digital beamforming only.

Due to the unreliable nature of the wireless medium, both Wi-Fi 6 and 5G use automatic repeat request (ARQ) that automatically initiates a request to retransmit frames received with error. NR uses hybrid ARQ retransmission with incremental redundancy where the base station retransmits the data and the device combines the soft information from multiple transmission attempts. Instead of retransmitting the same coded bits again, NR retransmits a different set of coded bits than the previous transmission providing extra information to the receiver at every retransmission. On the other hand, Wi-Fi 6 uses a simple, but less efficient, ARQ where the receiver completely discards the frames received with error and waits for the transmitter to retransmit the same coded bits again.

## 8.3 COMPARISON WITH WI-FI

While Wi-Fi has its origin in the personal computing industry, cellular technologies have their origins in the telephone industry. Because of this difference in their origins, Wi-Fi has remained the preferred connectivity technology for consumer electronics devices that are often used indoors, while telephone operators have more affinity to cellular and 5G. Historically, telephone operators, who are precursors to service providers and cellular operators, delivered services to the front door, leaving what happened inside the premises to the consumer. However, this mindset has been changing and cellular operators are also aspiring to deliver better quality indoor wireless internet services inside the customer premises. Whether this should be done by replacing Wi-Fi by extending cellular service indoors or by appropriately leveraging Wi-Fi has been a topic of debate within cellular and Wi-Fi communities. This section compares 5G, the latest cellular technology, and Wi-Fi 6, the latest Wi-Fi technology from various perspectives, including technology, security, ease of deployment, economics, and use cases, providing a viewpoint on this debate.

### 8.3.1 Technology

As discussed in Section 8.2, Wi-Fi 6 and 5G technologies have a lot in common. Both use OFDM and orthogonal frequency division multiple access (OFDMA) for PHY layer signaling and both can achieve multigigabit speeds and low latency. Wi-Fi 6 achieves a spectral efficiency of 60 bps/Hz, which exceeds the 5G requirement of 30 bps/Hz. Wi-Fi 6 is optimized for extremely dense environments, with a single Wi-Fi 6 access point capable of serving hundreds of clients concurrently. The trigger frame feature of Wi-Fi 6 enables scheduled access, like cellular, resulting in improved reliability of transmissions due to the elimination of collisions.

These two technologies handle network management differently. Wi-Fi uses unlicensed spectrum enabling anyone to deploy their own Wi-Fi network without needing any license. However, this can result in two adjacent networks being on the same channel impacting the overall network performance. This issue is mitigated in office and enterprise environments through network management. On the other hand, 5G networks typically are managed by operators and use a dedicated, licensed spectrum that requires subscription fees to access. This ensures that there is no cochannel interference.

Another distinction between cellular and Wi-Fi is with respect to the use of subscriber identity module (SIM). While Wi-Fi doesn't need a SIM card, cellular technologies use SIM cards to identify and authenticate devices. However, traditional SIM cards have become cumbersome for certain applications, including IoT. So, the cellular industry is moving to other options such as embedded SIM (eSIM) and soft SIM in lieu of the traditional SIM.

One of the areas where Wi-Fi falls short is mobility, as it is not specifically designed for high-speed mobility.

### 8.3.2 Security

Just as Wi-Fi security has improved over generations from WEP to WPA, subsequently to WPA2, and eventually to WPA3 as described in Chapter 4, cellular security has evolved with each generation. While the first-generation cellular systems didn't include any security, security was introduced in 2G. However, 2G only authenticates the terminal to the network and not vice versa. 3G offered greater security allowing mutual authentication between terminals and networks. 4G LTE further strengthened the encryption and authentication algorithms. 5G is expected to address security gaps found in 4G.

### 8.3.3 Ease of Deployment

Wi-Fi uses free unlicensed spectrum and does not require any complex backend infrastructure such as a packet core. It can be deployed in minutes without requiring a skilled technician. Cloud management has further simplified Wi-Fi deployment, making it as simple as plug and play. Now that the Wi-Fi calling feature is natively supported on most smartphones, Wi-Fi is a good alternative to deploying dual systems for calling.

### 8.3.4 Total Cost of Ownership

IP licensing costs associated with cellular technologies make cellular infrastructure and clients more expensive than their Wi-Fi counterparts. Unlike Wi-Fi, each new cellular generation is typically accompanied by new, and often expensive, spectrum. In addition, cellular services typically come with subscription fees paid to the network operator who owns the infrastructure and spectrum.

### 8.3.5 Use Cases

Owing to its lower cost to deploy, maintain, and scale, Wi-Fi will continue to remain the predominant technology to connect the growing number of data-hungry devices throughout the home and business, including PCs, tablets, smartphones, smart speakers, printers, streaming devices, TV sets, security cameras, thermostats, and other appliances.

Since 5G cellular has longer range, it will be used for mobile connections like smartphones, connected cars, smart city deployments, and large manufacturing operations.

Due to its ability to reach multigigabit speeds, service providers will offer broadband connectivity to homes through 5G in direct competition to cable and fiber. This will require mounting a 5G modem near a window since outdoor cellular signals have trouble penetrating buildings and the output of the 5G modem will be connected to a Wi-Fi router to provide in-home connectivity as done today. Connected cars is another use case where Wi-Fi and 5G may be used simultaneously: in-vehicle Wi-Fi is offered for users' devices, while the car itself connects to a 5G cellular network.

An emerging use case for wireless connectivity is mission-critical industrial applications such as automation using mobile robots or automated guided vehicles, and remote monitoring and diagnostics of industrial equipment. Private 5G networks are often promoted as the connectivity solution for this use case, where reliability, service quality, and security are key requirements. These networks are built similar to the standard public cellular networks, but utilize dedicated microtowers or small cell base stations to serve a particular premise. These networks can operate in either a licensed or unlicensed spectrum. However, operation in licensed spectrum requires obtaining a license from the local regulator or from an existing license holder, typically a mobile operator. Alternatively, operators can operate these private networks as a managed service with flat-rate pricing. Thus, the use of licensed spectrum limits the ease of deployment of private 5G networks, and makes them expensive to deploy and operate. These drawbacks are mitigated by operating the networks in unlicensed spectrum. However, the unlicensed spectrum introduces other requirements and restrictions—including listen-before-talk (LBT) for coexistence with other protocols such as Wi-Fi, and constraints on transmit power, power spectral density, and transmit duration—that can limit the efficacy of 5G-NR.

Both Wi-Fi and cellular technologies will continue to be strong complements to each other for the foreseeable future. In light of this, the Wireless Broadband Alliance is promoting OpenRoaming technology, which is built on the Passpoint (Hotspot 2.0) standard based on IEEE 802.11u, to allow devices move between Wi-Fi and cellular networks without interrupting service.

## 8.4 SUMMARY

**Table 8.1**

Cellular Evolution

| | *1G* | *2G* | *3G* | *4G* | *5G* |
|---|---|---|---|---|---|
| Year of introduction | 1980 | 1990 | 2000 | 2010 | 2020 |
| Data rate | 2 kbps | 64 kbps | 2 Mbps | 100 Mbps | 10 Gbps |
| Technologies | AMPS, TACS | GSM, IS-95, IS-136 | WCDMA, cdma2000 | LTE | NR |
| Service | Mobile Telephony | Digital Voice, SMS | Voice, Video, Data | Mobile Broadband | eMBB, mMTC, URLLC |
| Access technique | FDMA | TDMA/CDMA | CDMA | OFDMA | OFDMA |
| Duplex technique | FDD | FDD | FDD/TDD | FDD/TDD | FDD/TDD |
| Switching | Circuit | Circuit | Circuit/Packet | Packet | Packet |

As with Wi-Fi, cellular has gone through several generations of evolution over the last four decades as summarized in Table 8.1. 5G is the fifth and the latest generation of cellular technology. In addition to promising greatly enhanced speeds of around 10 Gbps, 5G is targeting very low latency as well as ability to connect a lot more devices. 5G operates in many different frequency bands – from 450 MHz to 52.6 GHz – to service a wide variety of use cases. Signal propagation and bandwidth availability at mmWave (24.25 – 52.5 GHz) is very different from signals below 6 GHz. While mmWave can achieve very high data rates by leveraging a very wide bandwidth, its range is limited because of the higher path loss at higher frequencies.

On the other hand, sub 6 GHz has good range, but the data rate is less due to the limited bandwidth.

Wi-Fi 6 and 5G technologies have a lot in common. Both use OFDM and OFDMA for PHY layer signaling and both can achieve multigigabit speeds and low latency. Given the favorable economics and high performance of Wi-Fi 6, Wi-Fi will remain a very attractive choice for indoor and business applications. While cellular has its origins outdoors, Wi-Fi and 5G are expected to coexist both indoors and outdoors for the foreseeable future.

## 8.5 REFERENCES

[1] Theodore S. Rappaport, *Wireless Communications: Principles and Practice*, Prentice Hall, 2001.

[2] Erik Dahlman, Stefan Parkvall, and Johan Skold, *5G NR: The Next Generation Wireless Access Technology*, Elsevier Academic Press, 2018.

# Chapter 9

# Conclusion

Wi-Fi is a household name that has become synonymous with access to the internet. With nearly 20 billion active Wi-Fi devices deployed globally, more than half the world's daily internet traffic is delivered through Wi-Fi. Wi-Fi 6 is the latest generation of Wi-Fi offering greater capacity, efficiency, and performance for advanced connectivity. Smartphones from Apple, Samsung, LG, and Huawei support Wi-Fi 6, and so do computers from Asus, Dell, HP, and Lenovo. Nearly 2 billion Wi-Fi 6 devices are expected to ship in 2021. Wi-Fi 6 networks are becoming more widely deployed in the homes and are also being steadily rolled out in the enterprise to ensure all networked devices perform at an optimal level.

Meanwhile, there is a global regulatory momentum in opening the 6-GHz band for unlicensed use. Following the landmark decision by the FCC, the United Kingdom, Europe, South Korea, Saudi Arabia, Canada, Chile, Brazil, the United Arab Emirates, Guatemala, Honduras, Morocco, Norway, Peru, and Costa Rica have made the 6-GHz spectrum available to Wi-Fi and many other countries are following their lead. Wi-Fi 6E will see rapid adoption in 2021 with more than 300 million devices entering the market.

Wi-Fi must continue to improve performance, increase spectrum efficiency, and, most importantly, make the user experience better to retain its prominence. The IEEE 802.11be extremely high throughput (EHT) task group has already started working on the next seventh generation Wi-Fi.

Wi-Fi 7 has ambitious goals and must meet aggressive requirements to meet our increasing connectivity needs. The IEEE is expected to ratify and publish the 802.11be amendment by 2024. Commercial equipment is expected around the same time, along with a certification program by the Wi-Fi Alliance to ensure interoperability.

Wi-Fi 7 will still use OFDMA, but with enhancements to make it more flexible and efficient. While Wi-Fi 6 is limited to 1024-QAM, Wi-Fi 7 will incorporate 4096-QAM. MU-MIMO will support 16 spatial streams, up from 8 in Wi-Fi 6. The maximum channel width is doubled to 320 MHz to fully exploit the 1.2 GHz of spectrum available in the recently added 6-GHz band. These new features of Wi Fi 7 effectively increase the maximum data rate to 46 Gbps. Apart from higher data rate, Wi-Fi 7 will also bring lower latency, as well as increased flexibility in using network and spectrum resources.

In summary, we are living in a golden age of connectivity. The speed, intensity, and breadth of communication is rising exponentially. Together with 5G, Wi-Fi will keep us connected and extend its reach to those among us who are still unconnected.

# List of Acronyms and Abbreviations

**AC** access category

**ACK** acknowledgment

**ADC** analog-to-digital converter

**AES** advanced encryption standard

**AFC** automatic frequency coordination

**AGC** automatic gain control

**AID** association identifier

**AIFS** arbitration interframe spacing

**AMPDU** aggregated MAC protocol data unit

**AMSDU** aggregate MAC service data unit

**AP** access point

**ASIC** application specific integrated circuit

**BA** blockack

**BAR** blockack request

**BCC** binary convolutional code

**BE** best effort

**BFRP** beamforming report poll

**BK** background

**BPSK** binary phase shift keying

**BSR** buffer status report

**BSS** basic service set

**BSSID** BSS identifier

**BTM** BSS transition management

**BW** bandwidth

**CAC** channel availability check

**CBF** compressed beamforming

**CCA** clear channel assessment

**CDMA** code division multiple access

**CFO** carrier frequency offset

**CP** cyclic prefix

**CQI** channel quality indication

**CRC** cyclic redundancy check

**CSA** channel switch announcement

**CSD** cyclic shift diversity

**CSMA-CA** carrier sense multiple access collision avoidance

**CSMA-CD** carrier sense multiple access collision detection

**CTS** clear to send

**DA** destination address

**DAC** digital-to-analog converter

**dB** decibel

**DCF** distributed coordination function

**DCM** dual carrier modulation

**DFS** dynamic frequency selection

**DH** Diffie Hellman

**DHCP** dynamic host configuration protocol

**DIFS** DCF interframe space

**DL** downlink

**DNS** domain name system

**DPP** device provision protocol

**DSL** digital subscriber line

**DSSS** direct sequence spread spectrum

**DTIM** delivery TIM

**EAP** extensible authentication protocol

**EAPOL** EAP over LAN

**ECDH** elliptic curve Diffie Hellman

**EDCA** enhanced distributed channel access

**EIRP** equivalent isotropically radiated power

**EMA** enhanced multi-BSSID advertisement

**EOSP** end-of-service period

**ETSI** European Telecommunications Standards Institute

**EU** European Union

**EVM** error vector magnitude

**FCC** Federal Communications Commission

**FCS** frame check sequence

**FFT** fast Fourier transform

**FHSS** frequency hopping spread spectrum

**FILS** fast initial link setup

**FT** fast BSS transition

**GHz** gigahertz

**GI** guard interval

**GMK** group master key

**GPS** global positioning system

**HD** high definition

**HE** high efficiency

**HT** high throughput

**IBSS** independent BSS

**IE** information element

**IEEE** Institute of Electrical and Electronic Engineers

**IFFT** inverse fast Fourier transform

**IGMP** internet group management protocol

**IoT** Internet of Things

**ISI** inter symbol interference

**ISP** internet service provider

**KRACK** key reinstallation attack

**L-LTF** legacy long training field

**L-SIG** legacy signal

**L-STF** legacy short training field

**LAN** local area network

**LDPC** low density parity check

**LNA** low noise amplifier

**LPI** low power indoor

**LTF** long training field

**MAC** medium access control

**Mbps** megabits per second

**MBSS** mesh BSS

**MCS** modulation and coding scheme

**MHz** megahertz

**MIC** message integrity check

**MIMO** multiple input multiple output

**MitM** man-in-the-middle

**MPDU** MAC protocol data unit

**MSDU** MAC service data unit

**MTU** maximum transmission unit

**MU** multiuser

**MU-MIMO** multiuser MIMO

**NAV** network allocation vector

**NDP** null data packet

**NDPA** null data packet announcement

**NOL** nonoccupancy list

**NSS** number of spatial streams

**OBSS** overlapping BSS

**OBSS PD** OBSS packet detect

**OCV** operating channel validation

**OFDM** orthogonal frequency division multiplexing

**OFDMA** orthogonal frequency division multiple access

**OKC** opportunistic key caching

**OMI** operating mode indication

**OWE** opportunistic wireless encryption

**PA** power amplifier

**PAPR** peak-to-average power ratio

**PCP** priority code point

**PE** packet extension

**PER** packet error rate

**PHY** physical

**PLCP** physical layer convergence procedure

**PMF** protected management frame

**PMK** pairwise master key

**PoE** power over Ethernet

**PPDU** physical layer conformance procedure protocol data unit

**ppm** parts per million

**PRI** pulse repetition interval

**PS** power save

**PSC** preferred scanning channel

**PSD** power spectral density

**PSDU** physical layer conformance procedure service data unit

**PSK** preshared key

**PSR** parameterized spatial reuse

**QAM** quadrature amplitude modulation

**QoS** quality of service

**QPSK** quadrature phase shift keying

**RA** receiver address

**RADIUS** remote authentication dial-in user service

**RAM** random access memory

**RBO** random backoff

**RF** radio frequency

**RNR** reduced neighbor report

**RRM** radio resource management

**RSN** robust secure network

**RSSI** receive signal strength indicator

**RTS** request to send

**RU** resource unit

**Rx** receive

**SA** source address

**SIFS** short interframe spacing

**SNR** signal-to-noise ratio

**SP** standard power

**SRG** spatial reuse group

**SSID** service set identifier

**SSN** starting sequence number

**STA** station

**STBC** space time block code

**STF** short training field

**SU** single user

**TA** transmitter address

**TB** trigger based

**TBTT** target beacon transmission time

**TCP** transmission control protocol

**TDMA** time division multiple access

**TID** traffic identifier

**TIM** traffic indication map

**TKIP** temporal key integrity protocol

**TSF** time synchronization function

**TU** time unit

**TWT** target wake time

**Tx** transmit

**TxBf** transmit beamforming

**TxOP** transmit opportunity

**U-NII** Unlicensed National Information Infrastructure

**UAPSD** unscheduled automatic power save delivery

**UL** uplink

**ULS** universal licensing system

**VAP** virtual access point

**VHT** very high throughput

**VI** video

**VO** voice

**VoIP** Voice over IP

**WDS** wireless distribution system

**WEP** wired equivalent privacy

**WFA** Wi-Fi Alliance

**WIDS** wireless intrusion detection system

**WIPS** wireless intrusion prevention system

**WLAN** wireless local area network

**WMM** Wi-Fi multimedia

**WNM** wireless network management

**WPA** Wi-Fi protected access

# About the Authors

**Susinder R. Gulasekaran** earned a bachelor's degree in electronics and communication engineering from College of Engineering, Guindy, Anna University and a Ph.D. degree in wireless communication from the Indian Institute of Science, Bangalore. His Ph.D. work was on designing space time block codes for colocated and distributed MIMO systems with low decoding complexity. He has contributed more than 20 publications in international journals/conferences and has 9 granted U.S. patents. Since 2008, he has worked on Wi-Fi products at various semiconductor and product companies including Atheros Communications, Qualcomm, Amazon Lab126, and Ruckus Networks. He currently serves as a director of engineering at Ruckus Networks, Commscope Inc. in Sunnyvale, California.

**Sundar G. Sankaran** is the VP of engineering at Verana Networks, a venture-backed startup building an innovative 5G radio access network solution. Sundar joined Verana from Ruckus Networks, where he, as a VP of engineering, led the Wi-Fi Access Point HW and SW engineering teams. Before joining Ruckus, he was a senior director of technology at Qualcomm Atheros and served as the overall engineering lead, with the responsibility to deliver silicon along with reference hardware and software, on multiple Wi-Fi chip programs. Prior to this, he has had stints at Intel, ArrayComm, and Infosys. He is a coinventor on 18 U.S. patents as well as several pending patents. Sundar earned a bachelor's degree in electronics and communication engineering from College of Engineering, Guindy, Anna University, and master's and Ph.D. degrees in electrical engineering from Virginia Tech.

# Index


ACK, CTS data rate, 37
acknowledgment frame, 12
aggregation, 23
  aggregate MPDU, 23
  aggregate MSDU, 23
  BA session, 25
    ADDBA request, 25
    ADDBA response, 25
    BA request (BAR), 25
    blockack (BA), 25
    DELBA, 26
    starting sequence number (SSN), 25
  delimiter, 24
  minimum MPDU start spacing, 24

beacon frame, 18
  beacon interval, 18
  BSS identifier (BSSID), 10, 18
  service set identifier (SSID), 18
  target beacon transmission time (TBTT), 18
  timing synchronization function (TSF), 31
broadcast frame, 18
BSS coloring, 79
  BSS color, 80
  BSS color change announcement, 81
  BSS color collision, 80
  inter-BSS, 80
  intra-BSS, 80
buffer management, 141
carrier sense multiple access collision avoidance (CSMA-CA), 17
  hidden node, 16, 17
  RTS-CTS frame exchange, 16
cellular technology, 197
  1G, 197
  2G, 198
  3G, 199
  4G, 199
  5G, 199, 200
    5G core network (5GCN), 200
    evolved packet core (EPC), 200
    millimeter wave (FR2), 201
    new radio (NR), 200
    nonstandalone access (NSA), 200
    standalone access (SA), 200
    sub-6 GHz (FR1), 201
channel bonding, 43
  non-HT duplicate PPDU format, 43
  primary channel, 43
  secondary channel, 43
channel delay spread, 37
channel frequencies, 146
  2.4-GHz band, 146
  5-GHz band, 148
  6-GHz band, 158, 159
channel switching, 34
  channel selection algorithm, 143
  channel switch announcement (CSA), 34
connection process, 21
  association identifier (AID), 19
  association request, 19
  association response, 19
  authentication request, 19
  authentication response, 19
  deauthenticate, 20

disassociation, 20
convolutional code, 37

data subcarrier allocation, 55
device provisioning protocol (DPP), 117
authentication phase, 118
QR code, 118
configuration phase, 118
configurator, 118
enrollee, 117
Wi-Fi easy connect, 117
distributed coordination function (DCF), 12, 14
clear channel assessment (CCA), 14
contention window, 14
DCF interframe space (DIFS), 14
extended interframe space (EIFS), 15
random backoff, 14
slot time, 14
virtual carrier sense, 15
downlink MU-MIMO, 47
HE DL MU-MIMO, 62, 70
null steering, 47
downlink MU-OFDMA, 60, 62, 64, 68
dual carrier modulation (DCM), 55
duration-based RTS-CTS, 59
dynamic channel bonding, 88

enhanced multi-BSSID advertisement (EMA), 86
multiple BSSID IE, 86
nontransmitted BSSID, 86
profile periodicity, 88
transmited BSSID, 86
enterprise Wi-Fi deployment, 178
capacity planning, 183, 185, 188
cell, 171
cell cluster, 181
channel reuse, 181
coverage planning, 183, 185
frequency reuse distance, 181
mitigating inter-AP interference, 189
hearing distance, 189
power over Ethernet (PoE), 178
upgrading existing deployment, 190

fast initial link setup (FILS), 164
EAP reauthentication protocol, 165
FILS connection steps, 166
Wi-Fi optimized connectivity experience (OCE), 164
fragmentation, 23
fragmentation threshold, 23
maximum transmission unit (MTU), 23
fragmentation and aggregation attacks (FragAttacks), 121
unauthenticated MAC header fields, 121

GI overhead, 55

half-duplex, 12
HE capability advertisement, 90, 92
HE capabilities IE, 53, 90
HE operation IE, 60, 90
high efficiency (HE) PHY, 51
home channel, 20

low density parity check (LDPC) code, 40

MAC frame format, 10
destination address, 10
duration, 11
frame check sequence (FCS), 10
MAC protocol data unit (MPDU), 10, 24
MAC service data unit (MSDU), 10, 23, 24
receiver address, 10
source address, 10
transmitter address, 10
MAC frame types, 12, 13
control frame format, 12
control frames, 12
data frames, 12
management frames, 12
information element, 12
medium access control (MAC) layer, 7
mesh Wi-Fi, 176, 177
802.11s mesh, 11, 176
WDS mesh, 176
Wi-Fi EasyMesh, 176
message integrity check (MIC), 109
midamble, 57
multi-TID AMPDU, 58

multicast traffic, 32
IGMP snooping, 32
multicast group, 32
multicast to unicast conversion, 32
multichannel man-in-the-middle, 119, 120
beacon protection, 121
operating channel validation, 120

network allocation vector (NAV), 11, 15
network architecture, 8
ad hoc, 8
basic service set (BSS), 8
mesh BSS, 8
wireless distribution system (WDS), 8

OpenRoaming technology, 206
Passpoint (Hotspot 2.0), 206
operating class, 158
operating mode indication (OMI), 89
opportunistic wireless encryption (OWE), 118, 119
Wi-Fi enhanced open, 119
orthogonal frequency division multiple access (OFDMA), 51
resource unit (RU), 51
RU locations, 62
RU sizes, 60, 61
orthogonal frequency division multiplexing (OFDM), 37
data subcarrier, 37
guard interval, 37
multicarrier modulation, 37
pilot subcarrier, 37
outdoor Wi-Fi deployment, 191
antenna types, 192
narrowbeam, 191
omnidirectional, 191
sectored, 191
point-to-point, 192
overlapping BSS, 79

packet processing speed, 142
small packet performance, 142
physical (PHY) layer, 7, 34
bandwidth, 35
coding, 35
coding rate, 35
data rate, 35
modulation, 35
constellation diagram, 35
spectral efficiency, 35
PLCP protocol data unit (PPDU), 24, 35
PLCP header, 35
PLCP preamble, 35
PLCP service data unit (PSDU), 24, 35
power save (PS), 26
delivery TIM (DTIM), 27
end of service period (EOSP), 30
power save notification, 27
PS exit, 28
PS poll, 29
STA power save algorithms, 145
traffic identification map (TIM), 27
unscheduled automatic power save delivery (UAPSD), 29
wireless network management sleep, 30
preamble puncturing, 88

quality of service (QoS), 21
access category (AC), 22
background (BK), 22
best effort (BE), 22
video (VI), 22
voice (VO), 22
arbitrary interframe spacing (AIFS), 22
enhanced distributed channel access (EDCA), 22
internal collision, 22
priority code point (PCP), 22
traffic identifier (TID), 22
Wi-Fi multimedia (WMM), 21

radio requirements, 129
device class, 134
frequency tolerance, 130
maximum Rx input level, 136
Rx blocker rejection, 136, 137
Rx sensitivity, 135
TB PPDU requirements, 135
Tx carrier leakage, 132
Tx error vector magnitude (EVM), 132–134

Tx spectral flatness, 132
Tx spectral mask, 131
rate adaptation algorithm, 138
received signal strength indicator (RSSI), 19
regulatory requirements, 146
dynamic frequency selection (DFS), 150
channel availability check (CAC), 151
channel closing Tx time, 152
channel move time, 152
ETSI off-channel CAC, 152
master device, 151
nonoccupancy list (NOL), 151
nonoccupancy period, 152
radar, 151
slave device, 151
ETSI power limits, 150
FCC power limits, 149
in-band power limit, 147
out-of-band emission limit, 147
power spectral density (PSD) limit, 147
regulatory requirements in 6 GHz, 167
ETSI low power indoor (LPI) operation, 170
ETSI very low power (VLP) operation, 170
FCC low power indoor (LPI) operation, 167
FCC standard power operation, 168
adaptive frequency coordination (AFC) system, 169
roaming, 32
BSS transition management (BTM), 33
BTM query, 33
BTM request, 33
fast BSS transition (FT) roaming, 112, 113
neighbor report, 33
opportunistic key caching (OKC), 112
radio resource management (RRM), 33
STA roaming algorithm, 144
rogue AP detection, 122
wireless intrusion detection system (WIDS), 123
wireless intrusion prevention system (WIPS), 123

scanning, 18
active scanning, 18
wildcard probe request, 18
off-channel scanning, 20
passive scanning, 18
scheduling algorithm, 140
short GI, 40
short interframe space (SIFS), 12
short slot time, 39
spatial multiplexing, 40
cyclic shift diversity (CSD), 41
number of spatial streams (NSS), 41
spatial multiplexing power save (SMPS), 31
dynamic SMPS, 31
static SMPS, 31
spatial reuse (SR), 81
dual NAV operation, 84
OBSS PD-based SR, 81
spatial reuse group OBSS PD, 84
PSR-based SR, 84

target wake time (TWT), 75
TWT agreement, 75
TWT service period (SP), 75
transmit beamforming (TxBf), 45
beamformee, 45
beamformer, 45
channel sounding, 45
compressed beamforming (CBF), 45
HE TxBf, 57
subcarrier grouping, 46

U-NII bands, 6
uplink MU, 63
buffer status report (BSR), 71
MU EDCA, 74
multi-STA BA, 67, 68
TB sounding sequence, 70
trigger frame format, 66
trigger frame types, 68, 69
basic trigger, 64
beamforming report poll (BFRP), 69
buffer status report poll (BSRP), 74
MU BAR, 68
MU RTS, 70
triggered response scheduling (TRS), 67
uplink MU-MIMO, 64
uplink MU-OFDMA, 64

virtual access point (VAP), 18

Wi-Fi 1 PPDU format, 36
Wi-Fi 2 PPDU format, 39
Wi-Fi 4 HT PPDU format, 43
Wi-Fi 5 VHT PPDU format, 45
Wi-Fi 6 BA lengths, 58, 59
Wi-Fi 6 HE PPDU formats, 52, 53
  extended range (ER) SU, 52
  GI and HE-LTF sizes, 54
  long OFDM symbol, 53
  multiuser (MU) PPDU, 52
  single user (SU) PPDU, 52
  trigger based (TB) PPDU, 52
Wi-Fi 6E, 157
  6 GHz operation information field, 162
  active scan in 6-GHz band, 159
  fast passive scanning, 161
    FILS discovery frame, 161
    preferred scanning channel (PSC), 161
  HE 6 GHz band capabilities IE, 162
  on channel tunneling (OCT), 163, 164
  out-of-band discovery, 160
    reduced neighbor report IE, 160
  security modes in 6-GHz band, 162
  short SSID, 159
  types of 6-GHz capable AP, 158
    6-GHz AP, 159
    6-GHz only AP, 159
Wi-Fi 7, 209
Wi-Fi certifications, 152
  Wi-Fi 6 mandatory features, 154
  Wi-Fi 6E mandatory features, 171
wired equivalent privacy (WEP), 104
  RC4 encryption, 104
  WEP authentication, 104
  WEP key, 104
  WEP vulnerabilities, 105
wireless network technologies, 195, 196
  WLAN, 195
  WPAN, 195
  WSN, 195
  WWAN, 195
wireless protected access (WPA), 106
  four-way handshake, 107
    EAPOL frames, 108
    group temporal key (GTK), 107
    pairwise master key (PMK), 107
    pairwise transient key (PTK), 107
  GTK rotation, 108
  Michael MIC, 109
  temporal key integrity protocol (TKIP), 106, 109
  WPA-Enterprise, 106, 108
    802.1X protocol, 108
    authentication server, 108
    authenticator, 108
    Diffie Hellman exchange, 108
    extensible authentication protocol (EAP), 108, 109
    remote authentication dial in user service (RADIUS), 108
    supplicant, 108
  WPA-preshared key (PSK), 106
    password, 106
    password-based key derivation function, 106
WPA2, 110
  advanced encryption standard (AES), 110
    AES-CCMP, 110, 111
    CBC-MAC MIC, 110, 111
  key reinstallation attack (KRACK), 114
  PMK caching, 111
  protected management frames (PMF), 112
    broadcast integrity protocol (BIP), 112
  robust secure network (RSN) IE, 110
  Wi-Fi protected setup (WPS), 113
    PIN method, 113
    push button method, 113, 114
WPA3, 114
  WPA3 encryption, 117
    AES GCMP, 117
  WPA3 transition Mode, 117
  WPA3-Enterprise, 116
  WPA3-simultaneous authentication of equals (SAE) authentication, 116
    dictionary attack, 116
    elliptic curve Diffie Hellman (ECDH) exchange, 116

## Artech House Mobile Communications Library

William Webb, Series Editor

*3G CDMA2000 Wireless System Engineering,* Samuel C. Yang

*3G Multimedia Network Services, Accounting, and User Profiles,* Freddy Ghys, Marcel Mampaey, Michel Smouts, and Arto Vaaraniemi

*5G New Radio: Beyond Mobile Broadband,* Amitav Mukherjee

*5G Spectrum and Standards,* Geoff Varrall

*802.11 WLANs and IP Networking: Security, QoS, and Mobility,* Anand R. Prasad and Neeli R. Prasad

*Achieving Interoperability in Critical IT and Communications Systems,* Robert I. Desourdis, Peter J. Rosamilia, Christopher P. Jacobson, James E. Sinclair, and James R. McClure

*Advances in 3G Enhanced Technologies for Wireless Communications,* Jiangzhou Wang and Tung-Sang Ng, editors

*Advances in Mobile Information Systems,* John Walker, editor

*Advances in Mobile Radio Access Networks,* Y. Jay Guo

*Artificial Intelligence in Wireless Communications,* Thomas W. Rondeau and Charles W. Bostian

*Broadband Wireless Access and Local Network: Mobile WiMax and WiFi,* Byeong Gi Lee and Sunghyun Choi

*CDMA for Wireless Personal Communications,* Ramjee Prasad

*CDMA RF System Engineering,* Samuel C. Yang

*CDMA Systems Capacity Engineering,* Kiseon Kim and Insoo Koo

*Cell Planning for Wireless Communications,* Manuel F. Cátedra and Jesús Pérez-Arriaga

*Cellular Communications: Worldwide Market Development,* Garry A. Garrard

*Cellular Mobile Systems Engineering,* Saleh Faruque

*Cognitive Radio Interoperability through Waveform Reconfiguration,* Leszek Lechowicz and Mieczyslaw M. Kokar

*Cognitive Radio Techniques: Spectrum Sensing, Interference Mitigation, and Localization,* Kandeepan Sithamparanathan and Andrea Giorgetti

*The Complete Wireless Communications Professional: A Guide for Engineers and Managers,* William Webb

*EDGE for Mobile Internet,* Emmanuel Seurre, Patrick Savelli, and Pierre-Jean Pietri

*Emerging Public Safety Wireless Communication Systems,* Robert I. Desourdis, Jr., et al.

*From LTE to LTE-Advanced Pro and 5G,* Moe Rahnema and Marcin Dryjanski

*The Future of Wireless Communications,* William Webb

*Geospatial Computing in Mobile Devices,* Ruizhi Chen and Robert Guinness

*GPRS for Mobile Internet,* Emmanuel Seurre, Patrick Savelli, and Pierre-Jean Pietri

*GSM and Personal Communications Handbook,* Siegmund M. Redl, Matthias K. Weber, and Malcolm W. Oliphant

*GSM Networks: Protocols, Terminology, and Implementation,* Gunnar Heine

*GSM System Engineering,* Asha Mehrotra

*Handbook of Land-Mobile Radio System Coverage,* Garry C. Hess

*Handbook of Mobile Radio Networks,* Sami Tabbane

*Handbook of Next-Generation Emergency Services,* Barbara Kemp and Bart Lovett

*High-Speed Wireless ATM and LANs,* Benny Bing

*Implementing Full Duplexing for 5G,* David B. Cruickshank

*In-Band Full-Duplex Wireless Systems Handbook,* Kenneth E. Kolodziej, Editor

*Inside Bluetooth Low Energy, Second Edition,* Naresh Gupta

*Interference Analysis and Reduction for Wireless Systems,* Peter Stavroulakis

*Interference and Resource Management in Heterogeneous Wireless Networks,* Jiandong Li, Min Sheng, Xijun Wang, and Hongguang Sun

*Internet Technologies for Fixed and Mobile Networks,* Toni Janevski

*Introduction to 3G Mobile Communications, Second Edition,* Juha Korhonen

*Introduction to 4G Mobile Communications,* Juha Korhonen

*Introduction to Communication Systems Simulation,* Maurice Schiff

*Introduction to Digital Professional Mobile Radio,* Hans-Peter A. Ketterling

*An Introduction to GSM,* Siegmund M. Redl, Matthias K. Weber, and Malcolm W. Oliphant

*Introduction to Mobile Communications Engineering,* José M. Hernando and F. Pérez-Fontán

*Introduction to OFDM Receiver Design and Simulation,* Y. J. Liu

*Introduction to Radio Propagation for Fixed and Mobile Communications,* John Doble

*Introduction to Wireless Local Loop, Broadband and Narrowband, Systems, Second Edition,* William Webb

*IS-136 TDMA Technology, Economics, and Services,* Lawrence Harte, Adrian Smith, and Charles A. Jacobs

*Location Management and Routing in Mobile Wireless Networks,* Amitava Mukherjee, Somprakash Bandyopadhyay, and Debashis Saha

*LTE Air Interface Protocols,* Mohammad T. Kawser

*Metro Ethernet Services for LTE Backhaul,* Roman Krzanowski

*Mobile Data Communications Systems,* Peter Wong and David Britland

*Mobile IP Technology for M-Business,* Mark Norris

*Mobile Satellite Communications,* Shingo Ohmori, Hiromitsu Wakana, and Seiichiro Kawase

*Mobile Telecommunications Standards: GSM, UMTS, TETRA, and ERMES,* Rudi Bekkers

*Mobile-to-Mobile Wireless Channels,* Alenka Zajić

*Mobile Telecommunications: Standards, Regulation, and Applications,* Rudi Bekkers and Jan Smits

*Multiantenna Digital Radio Transmission,* Massimiliano Martone

*Multiantenna Wireless Communications Systems,* Sergio Barbarossa

*Multi-Gigabit Microwave and Millimeter-Wave Wireless Communications,* Jonathan Wells

*Multiuser Detection in CDMA Mobile Terminals,* Piero Castoldi

*OFDMA for Broadband Wireless Access,* Slawomir Pietrzyk

*Practical Wireless Data Modem Design,* Jonathon Y. C. Cheah

*The Practitioner's Guide to Cellular IoT,* Cameron Kelly Coursey

*Prime Codes with Applications to CDMA Optical and Wireless Networks,* Guu-Chang Yang and Wing C. Kwong

*Quantitative Analysis of Cognitive Radio and Network Performance,* Preston Marshall

*QoS in Integrated 3G Networks,* Robert Lloyd-Evans

*Radio Resource Management for Wireless Networks,* Jens Zander and Seong-Lyun Kim

*Radiowave Propagation and Antennas for Personal Communications, Third Edition,* Kazimierz Siwiak and Yasaman Bahreini

*RDS: The Radio Data System,* Dietmar Kopitz and Bev Marks

*Resource Allocation in Hierarchical Cellular Systems,* Lauro Ortigoza-Guerrero and A. Hamid Aghvami

*RF and Baseband Techniques for Software-Defined Radio,* Peter B. Kenington

*RF and Microwave Circuit Design for Wireless Communications,* Lawrence E. Larson, editor

*Sample Rate Conversion in Software Configurable Radios,* Tim Hentschel

*Signal Processing Applications in CDMA Communications,* Hui Liu

*Signal Processing for RF Circuit Impairment Mitigation,* Xinping Huang, Zhiwen Zhu, and Henry Leung

*Smart Antenna Engineering,* Ahmed El Zooghby

*Software-Defined Radio for Engineers,* Travis F. Collins, Robin Getz, Di Pu, and Alexander M. Wyglinski

*Software Defined Radio for 3G,* Paul Burns

*Spread Spectrum CDMA Systems for Wireless Communications,* Savo G. Glisic and Branka Vucetic

*Technical Foundations of the IoT,* Boris Adryan, Dominik Obermaier, and Paul Fremantle

*Technologies and Systems for Access and Transport Networks,* Jan A. Audestad

*Third-Generation and Wideband HF Radio Communications,* Eric E. Johnson, Eric Koski, William N. Furman, Mark Jorgenson, and John Nieto

*Third Generation Wireless Systems, Volume 1: Post-Shannon Signal Architectures,* George M. Calhoun

*Traffic Analysis and Design of Wireless IP Networks,* Toni Janevski

*Transmission Systems Design Handbook for Wireless Networks,* Harvey Lehpamer

*UMTS and Mobile Computing,* Alexander Joseph Huber and Josef Franz Huber

*Understanding Cellular Radio,* William Webb

*Understanding Digital PCS: The TDMA Standard,* Cameron Kelly Coursey

*Understanding WAP: Wireless Applications, Devices, and Services,* Marcel van der Heijden and Marcus Taylor, editors

*Universal Wireless Personal Communications,* Ramjee Prasad

*WCDMA: Towards IP Mobility and Mobile Internet,* Tero Ojanperä and Ramjee Prasad, editors

*Wi-Fi 6: Protocol and Network,* Susinder R. Gulasekaran and Sundar G. Sankaran

*Wireless Communications in Developing Countries: Cellular and Satellite Systems,* Rachael E. Schwartz

*Wireless Communications Evolution to 3G and Beyond,* Saad Z. Asif

*Wireless Intelligent Networking,* Gerry Christensen, Paul G. Florack and Robert Duncan

*Wireless LAN Standards and Applications,* Asunción Santamaría and Francisco J. López-Hernández, editors

*Wireless Sensor and Ad Hoc Networks Under Diversified Network Scenarios,* Subir Kumar Sarkar

*Wireless Technician's Handbook, Second Edition,* Andrew Miceli

For further information on these and other Artech House titles, including previously considered out-of-print books now available through our In-Print-Forever® (IPF®) program, contact:

Artech House
685 Canton Street
Norwood, MA 02062
Phone: 781-769-9750
Fax: 781-769-6334
e-mail: artech@artechhouse.com

Artech House
16 Sussex Street
London SW1V 4RW UK
Phone: +44 (0)20 7596-8750
Fax: +44 (0)20 7630-0166
e-mail: artech-uk@artechhouse.com

Find us on the World Wide Web at: www.artechhouse.com